Nature, Empire, and Nation

Nature, Empire, and Nation
*Explorations of the History of Science
in the Iberian World*

Jorge Cañizares-Esguerra

STANFORD
UNIVERSITY
PRESS

Stanford,
California

2006

Stanford University Press
Stanford, California

This book has been published with the assistance of a University
Cooperative Society Subvention Grant awarded by The University
of Texas at Austin.

Library of Congress Cataloging-in-Publication Data

Cañizares-Esguerra, Jorge.
   Nature, empire, and nation : explorations of the history of
science in the Iberian world / Jorge Cañizares-Esguerra.
      p.   cm.
Includes bibliographical references and index.
   ISBN-13: 978-0-8047-5543-6 (acid-free paper)
   ISBN-10: 0-8047-5543-4 (acid-free paper)
   ISBN-13: 978-0-8047-5544-3 (pbk. : acid-free paper)
   ISBN-10: 0-8047-5544-2 (pbk. : acid-free paper)
   1. Science—Latin America—History—19th century.
2. Spain—Colonies—Latin America.   3. Science—Spain—
History—19th century.   4. Spain—Politics and govern-
ment—18th century.   5. Creoles—Latin America—Intellectual
life—17th century.   I. Title.
Q127.L29 C36   2006
509.171'246—dc22   2006005163

Typeset by BookMatters in 10/13 Electra

For Sandra, always

Contents

Illustrations

Illustrations

Acknowledgments

An earlier version of Chapter 1 appeared in the journal *Isis*. I am grateful to the University of Chicago Press for allowing me to reproduce it here and thankful for the valuable insights of Londa Schiebinger and Bernie Lightman.

An earlier version of Chapter 2 appeared in *Perspectives on Science*. MIT Press has kindly allowed me to reproduce it. Mordechai Feingold and Victor Navarro Brotóns read it, and their comments improved the final result. For bibliographic guidance, I thank Antonio Barrera and Alison Sandman. Invitations to the History of Science Colloquium at UCLA (Mary Terrall) and the workshop "Science Across Cultures: Historical and Philosophical Perspectives" at Princeton University (Helen Tilley and Daniel Garber) also contributed to improving the essay.

An earlier version of Chapter 3 was published in *The Cambridge History of Science*, volume 4: *Eighteenth-Century Science*, edited by Roy Porter, David C. Lindberg, and Ronald L. Numbers (New York: Cambridge University Press, 2003). I am grateful to Cambridge University Press for allowing it to be reprinted here. The comments of Marcos Cueto, Felipe Fernández-Armesto, Thomas Glick, Antonio Lafuente, and Roy Porter were most helpful.

An earlier version of Chapter 4 appeared in the *American Historical Review*, and I thank the History Cooperative for permitting it to be reprinted here. The essay was written with the generous intellectual help of Tom Broman, Joyce E. Chaplain, Victor Hilts, R. Douglas Cope, Sharon Farmer, Felipe Fernández-Armesto, John Freed, Paul Freedman, Anthony Grafton, Steve Hackel, Sabine MacCormack, John Markoff, Susan Niles, Karen Ordahl Kupperman, Pilar Ponce Leiva, Mohamad Tavakoli-Targi, and five *AHR* anonymous reviewers. I also benefited from the insight of audiences who heard different drafts of this chapter in presentations or conferences sponsored by the Consejo Superior de Investigaciones Científicas (Madrid); Illinois State University; the Society for the Social History of Medicine; the John Carter Brown Library at Brown University; the Forum on European Expansion and Global Interaction; the History of Science Society; and the

Conference on Latin American History. Research for this essay was undertaken with the help of a National Endowment for the Humanities Fellowship at the John Carter Brown Library (1997–98) and two small research grants from Illinois State University.

An earlier version of Chapter 5 was published in *Eighteenth-Century Thought,* under the learned and gentle editorial hand of James J. Buickerood. The essay originated as a result of a kind invitation by Antonio Feros and John H. Elliott to present an earlier version in Spanish at the seminar "La evolución histórica de la España moderna: Éxitos y fracasos" in Soria, Spain, in July 2001, generously financed by the Fundación Duques de Soria. Sharp criticism by two anonymous reviewers of the original Spanish version forced me to clarify my arguments. Antonio Feros also offered pointed commentary and bibliographical help as I composed a new draft in English. Finally, the members of the Workshop on Early Modern Studies at the University of Chicago read and commented on the essay and helped me rethink my conclusions.

An earlier version of Chapter 6 appeared in *Colonial Botany: Science, Commerce, and Politics in the Early Modern World,* edited by Londa Schiebinger and Claudia Swan (Philadelphia: University of Pennsylvania Press, 2005). The University of Pennsylvania Press has kindly allowed it to be reproduced here. Londa Schiebinger helped me improve the first drafts, and I received friendly and constructive feedback from audiences at the King Juan Carlos Center at New York University (Antonio Feros); Stanford University (Zephyr Frank); and the German Historical Institute (Andreas Daum).

Chapter 7 was written during my tenure as a fellow of the Charles Warren Center for Studies in American History at Harvard University. Pat Denault, Joyce Chaplin, Charles Rosenberg, and all the other 2001–2 fellows went out of their way to make the academic year there fun and productive. Mauricio Tenorio-Trillo brought to my attention a serious structural error in an earlier version of the essay. Charles Hale and Ann Bermingham offered sympathetic and useful readings. Invitations by Peter Miller to the Bard Graduate Center in New York City, by David Tavares to Bard College, and by history graduate students to Yale University and to the University of Texas at Austin helped me to sharpen the argument as I presented earlier versions of the essay in public.

The chapters have been edited to eliminate redundancies owing to the separate publication of the original essays. At Stanford University Press, Norris Pope and Anna Eberhard Friedlander carried this book through the process of publication, while Peter Dreyer expunged errors.

Finally, I am grateful to the University of Texas at Austin for a University Cooperative Society Subvention Grant that facilitated this publication.

Nature, Empire, and Nation

Introduction

This book is a collection of essays, written over the course of ten years, on the history of science in the early modern and nineteenth-century Iberian world. "Science" is a problematic word when it comes to describing early modern learned perceptions of nature. The term only began to take on the sense in which we use it now in the early eighteenth century, and it does not do justice to the ideas and practices of the early modern world, for many of the historical actors the reader will encounter in these pages were alchemists, not chemists, curiosity collectors, not biologists, astrologers, not astronomers, landscape artists, not geologists, and polymaths rather than specialists. "Iberia" is also a somewhat misleading term: the essays here are about the viceroyalties that made up the early modern Spanish empire and the new nineteenth-century national polities that succeeded them. This was a world with many conflicting interpretations of nature, and I have focused on only a handful here. On the one hand, I explore the metropolitan, instrumental imperial perspectives. Whether it is the seventeenth or the nineteenth century, I identify styles of science that were imperial and had a long-lasting influence on other European empires, the narrative of the so-called Scientific Revolution notwithstanding. On the other hand, I explore interpretations of nature that were "patriotic"—that is, that sought to defend the Spanish American viceroyalties or new emerging nations from European innuendos.

The first essay, "Chivalric Epistemology and Patriotic Narratives: Iberian Colonial Science," offers a quick overview of these two traditions in colonial Spanish America, laying out the ground for the other essays. It suggests that the epistemology colored by "chivalric," crusading values that emerged in the crucible of the Portuguese and Spanish expansions to Africa, India, and the New World was also influential in the early modern English and French empires.

Chivalric epistemology is also the focus of the second essay, "The Colonial Iberian Roots of the Scientific Revolution," which explores the Iberian origins of Francis Bacon's ideas in his unfinished utopian fable *The New Atlantis* (1624; published posthumously in 1627). It argues that historians of science have overlooked the Iberian colonial origins of key ideas associated with the seventeenth-century changes in natural philosophy that revolutionized the early modern world. Drawing upon the research of Spanish historians of science, the essay exposes the biases of Anglo-American scholarship that has remained blind to the Iberian origins of modernity. Locating those origins in northern Europe has not simply been the product of cultural biases that have marginalized developments in the Spanish empire as "backward"; it has also been the result of what disciplines historians of science have chosen to study to construct their narratives. Until recently, it was mathematics and physics that monopolized all the attention; the Scientific Revolution was seen as the outcome of the triumphs of Copernican heliocentrism and the new mechanical philosophy of Galileo, Descartes, and Newton. Historians are abandoning this narrative, however, and paying closer attention to the roles of empire, commerce, and collecting. Disciplines like cartography, mapmaking, and natural history are therefore no longer marginal to the narratives of the origins of scientific modernity. Yet despite this switch, historians of science have not been flocking to study the Iberian empires. The essay suggests that this has to do in part with the manuscript, scribal culture of the Spanish empire, which fostered a tradition of state secrecy that kept most of its investigations into nature and technology in archives, unavailable to other Europeans and the collective historical memory.

The third essay, "From Baroque to Modern Colonial Science," offers a general overview of the practices of natural philosophy in colonial Spanish America. The early modern composite monarchies endowed the societies of the New World with remarkable degrees of political and economic autonomy. As in Europe, a Neoplatonic tradition of science developed to bolster the local hierarchical polities. This baroque tradition came under attack by the early eighteenth century, however, when the Spanish empire sought to reform the state and transform its New World viceroyalties into modern colonies. This resulted in the development of new traditions of science, which in turn led to profound cultural changes. Yet despite the changes, nature was deployed under both the "baroque" and the modern colonial regimes to support patriotic agendas within the New World polities.

The fourth essay, "New World, New Stars: Patriotic Astrology and the Invention of Amerindian and Creole Bodies in Colonial Spanish America, 1600–1650," describes in greater detail how patriotism colored the practice of astrology in the New World. It also shows that the Creole learned elites in

medicine used viceroyalties
to defend by creoles

Spanish America were willing to transform Galenic and Hippocratic medicine to defend their viceroyalties, which they considered to be full-fledged "kingdoms," from foreign innuendos. In the process, they created "modern" understandings of the racialized body.

Patriotism was not the exclusive monopoly of Spanish Americans, of course, and in the fifth essay, "Eighteenth-Century Spanish Political Economy: Epistemology and Decline," I argue that it was a central intellectual force in Spain as well. As eighteenth-century Spanish political economists grappled with the causes of imperial decline, they had to contend with northern European theories. Some political economists embraced these theories, although they were suspicious of the reliability of foreign observers. Others, however, marked sharp epistemological differences with the northern Europeans and insisted that the causes of Spain's decline could only be explained through careful archival research, not by theories based on first principles of human nature. This anti-Cartesian patriotic tradition developed an epistemology of the social sciences remarkably similar to that introduced by such critics of the French Revolution and the philosophes as Edmund Burke (1727–97).

The sixth essay, "How Derivative Was Humboldt? Microcosmic Narratives in Early Modern Spanish America and the (Other) Origins of Humboldt's Ecological Sensibilities," continues the study of patriotic interpretations of nature in the Spanish empire but does so with a twist. It argues that the history of ecological thought has ignored the Spanish American roots of Alexander von Humboldt's theories of biogeography. Humboldt (1769–1859) has been hailed as the inventor of theories of plant ecology and biodistribution that culminated in his maps of the microecological niches of the Andes. Historians of science have, however, made little of the fact that Humboldt learned to read the Andes as a natural laboratory in which to study plant distribution from local colonial intellectuals. Local naturalists understood the Andes to be a microcosm endowed with all the climates and thus all the fauna and flora of the world. In their imagination, the Iberian viceroyalties were the world writ small: polities poised on the verge of becoming commercial empires.

The final essay, "Landscapes and Identities: Mexico, 1850–1900," argues that nineteenth-century Mexican scientific artists trained in the new sciences of geology and meteorology produced remarkable representations of local landscapes that were in fact historical allegories of the nation. These intellectuals considered the nation to be constituted of layers of various historical memories to be read in the landscape, including both the indigenous

5

precolonial past and Spanish colonial and religious legacies. Representations of the landscape were also used to postulate a modern synthesis for the nation, a way out of the secular racial, economic, and political crises that were the bane of nineteenth-century Mexico. These interpretations of the landscape were also patriotic, for they set out to undermine the foreign views of Mexico, which ever since Humboldt had seen the land as a collection of plant and animal types. Foreigners obsessed with Humboldtian biodistribution drew landscapes of "empty" lands. To counter this imperial tradition, Mexican scholars transformed landscapes into cityscapes, firmly rooting Mexico in ancient indigenous and Mediterranean urban traditions.

Chivalric Espistemology and Patriotic Narratives
Iberian Colonial Science

Of all the modern European powers, Spain and Portugal have by far the longest colonial record, beginning in the fourteenth century with the settlement of the Canary Islands and ending in the late twentieth century with the protracted processes of decolonization of Angola, Macao, East Timor, and the Spanish Sahara. Yet despite this record, our knowledge of the role of science and technology in the making, spread, and survival of these colonial empires is spotty and limited.[1] Spanish historians, in particular, have shed light on the history of cartography, metallurgy, and natural history in Spanish America in the sixteenth and eighteenth centuries.[2] Beyond these two centuries and this region, however, we know little of Iberian colonial science and technology. Within Spanish America itself, our knowledge of the history of the colonial science of the seventeenth and nineteenth centuries is sketchy and blurry, to say the least. The so-called decline of Spain has made historians of science focus mostly on the sixteenth-century apogee of the empire and its eighteenth-century Bourbon revival, to the detriment of the long seventeenth century, which was allegedly characterized by forms of decadent, baroque scholarship. The nineteenth century, too, has been marginalized, largely owing to the fact that Spain lost most of its colonial territories in the New World in the 1820s.[3] But nineteenth-century Spain continued to be a formidable empire, with far-flung colonies in America, Africa, and the Pacific. Cuba and Puerto Rico remained central to the economy and politics of metropolitan Spain.[4]

The gaps in our knowledge are considerable, even in areas that have attracted substantial attention, as in the case of the history of botany and empire. The collection of plants to secure new monopolies and trade was accomplished by at least four means: metropolitan expeditions, the patriotic impulse of provincial clerical elites within the composite monarchy that was the empire, the private initiative of entrepreneurial settlers and merchants, and coordinated regional, and even continental, campaigns to gather information using the empire's bureaucracies and the vast network of municipal

7

authorities (the *relaciones geográficas*). Our knowledge of each of these activities and institutions varies considerably.

The scholarship on the botanical expedition in Mexico of the royal physician Francisco Hernández (1517–87), for example, is vast.[5] At a time when Europeans had inherited from Dioscorides, Theophrastus, and Arab naturalists knowledge of no more than 600 species of plants, Hernández put together some fifteen volumes of writings and illustrations of 3,000 new plants. Hernández used the fledgling network of colonial hospitals in Mexico to test the medical properties of hundreds of these plants. With the help of local Nahua intellectuals trained in Latin and classical sources by Franciscan missionaries, Hernández tapped into centuries of Nahua (Aztec) botanical and medical scholarship. Historians have, for example, identified in the extant images of plants assembled by the expedition (the eleven volumes of images put together by Hernández were lost in the great fire of El Escorial in the late seventeenth century) traces of Nahua iconographic (pictographic, hieroglyphic) conventions.[6] Less well explored, however, is the fact that Hernández's written descriptions of plants in Latin are remarkably similar to those offered in the *Libellus de medicinalibus Indorum herbis*, an earlier illustrated herbal by two Nahua intellectuals, Juan Badiano and Martín de la Cruz.[7]

That Badiano and Cruz wrote an herbal in Latin in Mexico City as early as 1552 speaks to the traditions of Nahua botanical knowledge into which the Spaniards tapped, as well as to the rapid processes of global cultural mongrelization and hybridization that the conquest sparked. The cases of Badiano and Cruz's and Hernández's natural histories should teach historians of imperial science to avoid reductive terms such as "European" and "indigenous," both based on reified and dichotomous notions of identity.[8]

But Badiano and Cruz's herbal also speaks to the rich traditions of local, colonial botanical and natural history writing supported by religious orders (Badiano and Cruz were trained by, and wrote for, the Franciscans). Steven J. Harris has explored some of these writings, particularly those produced by Creole (Spanish American) Jesuits. There is no denying that this tradition had a utilitarian, mercantile goal (identifying medicines for infirmaries and hospitals, or, as in the case of the Jesuits, securing worldwide monopolies). By and large, however, local clerics wrote with purposes other than commerce and trade in mind. They sought to create patriotic genealogies for the new American viceroyalties as members of the Spanish composite monarchy. The clergy also turned to the surrounding landscape to secure providential, moral narratives.[9] There was, to be sure, no novelty to this tradition. Members of the local intelligentsia of every kingdom that constituted

the Spanish empire, from Naples, Sicily, and Aragon to México, were heavily invested in developing patriotic surveys of local material and spiritual
resources, including not only natural histories but also chorographies and
the ubiquitous hagiographies.[10] Thus it would be an error to exaggerate the
distinction between European "provinces" and New World "colonies" in the
early modern composite monarchy.[11]

In addition to clerics engaged in patriotic and moral campaigns, there
were enterprising settlers and merchants who upon arrival set out to identify
plants that could be sold in Europe. It is often wrongly assumed that the
Iberians either established local clerical botanical traditions or dispatched
expeditions to collect botanical information, while the English mostly left
systematic botanical work to merchants and other enterprising individuals. The large scale of private botanical initiatives in the Spanish empire is
thankfully, however, being clarified by the ongoing archival researches of
Antonio Barrera and Paula De Vos. Both Barrera and De Vos have exposed
a world of enterprising settlers and merchants who claimed to have New
World plants with extraordinary curative powers. De Vos has found shipment after shipment of curiosities in the eighteenth century.[12] Barrera has
demonstrated that the surge of claims and counterclaims as to the curative
virtues of plants forced the Spanish Crown to create institutions such as
the Casa de Contratación to apportion credit, settle disputes, and hand out
patents in the sixteenth century. Apothecaries and physicians back in Spain
sought to systematize this new knowledge created by settlers and merchants.
Nicolás Monardes is well known for his efforts to survey these new drugs,
which had been flooding the markets of Seville since the early sixteenth
century.[13] Yet despite the efforts of scholars such as Barrera and De Vos, we
know very little about this market for botanical goods in the Spanish empire.
A history of the trade in plants and dyes, such as sarsaparilla, cochineal, ipe-
cacuanha, guaiacum, cacao, jalap root, vanilla, and the precious febrifuge
cinchona, over which Iberians had held very lucrative monopolies since the
sixteenth century, has yet to be written.[14]

However, there was yet another mechanism used in the Spanish empire
to gather knowledge of colonial botanical resources. Scholars have studied
the campaigns known as *relaciones geográficas*, in which thousands of local
authorities answered detailed questionnaires sent out by the Crown about
the nature of the local resources, in some detail. The answers now fill
dozens of volumes in archives on both sides of the Atlantic. Of all the traditions I have described above (the others being expeditions, clerical natural
histories, and entrepreneurial settlers), this is perhaps the only one that was

unique to the Spanish empire. The *relaciones* were sweeping, continentwide surveys with pretensions to all-encompassing knowledge, launched by rickety premodern bureaucracies with no standing armies at their disposal. It was the vast network of municipal authorities, both indigenous and Creole, that mobilized resources to reply to the Crown questionnaires, thus leading to remarkable variations.[15]

As this quick overview of Spanish colonial botanical efforts should demonstrate, there are big gaps in our knowledge. To be sure, much remains to be done in other sciences, such as cartography and metallurgy, that, like natural history, were born out of a colonial context.[16] There are also areas wholly unexplored. Let me turn to two in the remainder of this chapter. One is the elucidation of a model of imperial science, popularized by early modern Spain, in which chivalric, gendered values colored the pursuit of knowledge. This model had a wide impact on early modern British and French colonialism as well. The second area is the different aspects specific colonial sciences took on when practiced with imperial or patriotic aims in mind.

The organization and pursuit of knowledge in chivalric terms, that is, the cosmographer as knight, or the knight as cosmographer, was a hallmark of the Portuguese and Spanish fifteenth- and sixteenth-century colonial expansion. A quick glance at the sixteenth-century Iberian treatises of cosmography so much admired by the English shows that the Iberians saw knowledge gathering as an expansion of chivalric virtues.[17] This model made a profound impression on Elizabethan proponents of English colonial expansion, such as Sir Walter Raleigh and the community of mathematicians, cosmographers, and alchemists who gathered around him, including Thomas Harriot, John Dee, and Lawrence Kemys. It was the learned Kemys who, in his account of his second trip to Guyana in 1596 to recover samples of gold, insisted that the Orinoco should be named the "Raleana" just as the Amazon had been named the "Orellana" after its Spanish discoverer, Francisco de Orellana. This chivalric model also had lasting influence on Captain John Smith and the colonization of Virginia. Smith liked to present himself as a knight cosmographer, wielder of both the sword and the compass. In a map of his *Advertisements for the unexperienced planters of New-England* (London, 1631), Smith appears as a fully armored knight, standing right next to a globe.

These images resemble those of Amerigo Vespucci and Fernão de Magalhães (Ferdinand Magellan) that the Flemish artist Jan van der Straet (1523–1605) (aka Johannes Stradanus) bequeathed to posterity in his *Americae retectio* (ca. 1589) (fig. 1.1). The third engraving (bottom left) has Vespucci

FIG. 1.1. Three heroes cross the ocean, guided by Providence. Drawing by Johannes Stradanus (Jan van der Straet), engraved by Adrainus Collaert and printed by Ioannes Galle in *Americae retectio* (ca. 1589). From Theodore de Bry et al., *Americae pars quarta* (Frankfurt a/M, 1594). The originals (inverted) are reproduced in Straet, *New Discoveries* (1953).

making astronomical observations with a quadrant. Next to him is a banner bearing the cross, a reminder that Vespucci first described the constellation of the Southern Cross. A broken mast reminds the viewer that the knight-cosmographer has survived a tempest. The final image (bottom right) portrays Magellan crossing the straits (represented here by a Patagonian giant swallowing an arrow, to the right, and the Land of Fire, to the left). Magellan is depicted as a knight clad in full armor who charts the heavens by means of an armillary sphere, a lodestone, and a compass, led by Apollo, the sun god, carrying a lyre, and aided by Aeolus, god of the winds. This gendered, epic, and markedly aggressive notion of the role of knowledge in the expansion of empire is one of many traditions inherited by northern Europe from sixteenth-century Portuguese and Spanish colonialism.

To understand the different ways in which science was deployed in the Spanish empire, we should not assume that the sprawling "empire" Spain created in the early modern period resembled that built by England in the nineteenth and twentieth centuries. From the beginning, the colonies Spain acquired in the New World were considered by the settlers to be "kingdoms," part of the larger composite monarchy that was Spain.[18] This status of "kingdoms" was not merely symbolic or misleadingly rhetorical. Spanish America developed as a typical ancien régime society, in which corporate privileges and social estates took on an added racial dimension (i.e., Amerindians as peasants; blacks as slaves). Until the eighteenth century, the Creole elites of these colonial viceroyalties enjoyed considerable autonomy and thus developed strong local identities and patriotic narratives. Thus it was typical that for every "imperial" version of a science that arrived in America a local, "colonial" version emerged. I have discussed astronomy and astrology in the seventeenth century elsewhere. Whereas European scholars developed a largely negative view of the effect that the stars of the Southern Hemisphere had on the New World's peoples, fauna, and flora, Creole intellectuals in Spanish America begged to differ. The latter not only challenged European estimates of the size and number of the newly discovered stars and constellations but also developed patriotic and providential alternative astrological readings.[19] I want here to describe briefly the case of eighteenth-century botany.

Having been soundly defeated by the British in the Seven Years' War (1756–63), the Spanish Bourbons sought to introduce aggressive economic, administrative, and cultural reforms in every corner of their far-flung empire. Determined to transform the Spanish viceroyalties into colonies, the Spanish Bourbons turned to the new sciences. The Spanish empire had long been losing territories in the New World, along with status and prestige, to other European powers. Some Spanish intellectuals maintained that the loss of territories began with losses in the struggle over naming, surveying, and remembering.[20] The writing of histories of "discovery" and colonization and the launching of cartographic and botanical expeditions therefore became priorities, and some twenty-five such expeditions visited the New World. Naturalists sought to benefit the economy by identifying new products (dyes, spices, woods, gums, pharmaceuticals) or alternatives to already profitable staples from Asia. Spanish botanical expeditions to the Andes, for example, put a premium on finding species of cloves and cinnamon to challenge the British and Dutch monopolies in the East Indies.[21]

These plans to grow cloves, cinnamon, and other spices in the New World to break the Dutch and British monopolies came to nought, but the cultural transformation the Bourbon botanists brought about was profound. Botany became new cultural capital in the form of providential idioms and discourses highlighting the untapped economic potential of each viceroyalty. Projects designed to turn local societies into subordinate appendages of a new, revitalized modern empire unwittingly offered ideological tools that allowed those communities to think of themselves, literally, as middle kingdoms. Like their counterparts in contemporary Qing China, Creole intellectuals came to think of their local polities as the center(s) of the world. Spanish American naturalists came to represent each *patria* as a microcosm wondrously poised to become a trade emporium.[22] Mauricio Nieto Olarte has argued in his book *Remedios para el imperio* (2000) that Creole traditions of botany in every respect resembled the colonial model introduced in the eighteenth century by metropolitan naturalists. My interpretation differs from that of Nieto Olarte in that I argue that the mercantilist, pragmatic, imperial goals of metropolitan expeditions were transformed in the Spanish American viceroyalties into utopian patriotic accounts of the landscape and nature. Once the imperial science of Linnaean botany arrived in the "tropics," it took on a life of its own, and it was eventually deployed by local patriot-naturalists to undermine the very goals that Linnaean natural history had set out to accomplish in Spanish America, namely, to revamp and strengthen the empire.

Studies of Iberian traditions of colonial science have been largely limited to the history of early modern Spanish America. We know relatively little of other centuries and regions of the world (to say nothing of the Portuguese side of the story). Notwithstanding these limitations, or rather scholarly challenges, it should be clear, however, that any understanding of European traditions of colonial science needs to come to grips with long-term patterns that first emerged in the tumultuous multicultural world of the early modern Iberian empires.

The Colonial Iberian Roots
of the Scientific Revolution

Some readers may be familiar with the frontispiece to Francis Bacon's *Instauratio magna* (1620), which shows a ship sailing through the pillars of Hercules (fig. 2.1), standing for the voyage of empirical and experimental discovery of nature's secrets on which Europeans embarked as soon as they put the authority of ancient texts behind them. The introductory engraving in Jan van der Straet's *Nova reperta* (ca. 1580), a collection of thirteen illustrations of key early modern "discoveries," also highlights the importance of new technologies such as the printing press, the clock, and the cannon (fig. 2.2).

Paradoxically, these two illustrations have come to be associated first with a "Protestant" and later with an "Enlightenment" narrative of modernity, which purposefully obscures the role of Catholic Iberia in the so-called Scientific Revolution. I say paradoxically because these two illustrations either drew on Iberian motifs of discovery or sought to capture Iberian contributions to knowledge.

Take Francis Bacon's motif of sailing through the pillars of Hercules to signify the triumph of the moderns over the ancients, for example, which was in fact a sixteenth-century Spanish export. As José Antonio Maravall persuasively argued some forty years ago, it was in the Iberian Peninsula, and particularly Spain, that intellectuals first developed a sense that the moderns had superseded the ancients. The discovery of hitherto unknown patterns of oceanic wind currents, the development of new vessels, and the mastery by sailors of new techniques to find their bearings in the open sea led in the fifteenth century to a growing realization that the cosmographies inherited from the ancients were wrong. By the early sixteenth century, Iberians had discovered that the Indian Ocean was not an inland sea, as Ptolemy maintained, and that there was a whole new world in the middle of the Atlantic. Numerous peoples who had clearly been unknown to the ancients were discovered in Africa and the Americas. The new empires that the Portuguese and Spaniards put together, which encompassed peoples and bureaucracies

FIG. 2.1. Frontispiece to Francis Bacon's *Instauratio magna* (London, 1620). New York Public Library.

FIG. 2.2. Introductory image in Jan van der Straet's *Nova reperta*, engraved by Philippe Galle (ca. 1580). The Burndy Library, Dibner Institute for the History of Science and Technology, Cambridge, Massachusetts.

on four different continents, were wonders of political engineering and military prowess that dwarfed anything Rome had ever accomplished. No wonder that when looking for motifs to capture the deeds of his new Holy Roman Empire, Charles V (1500–1558) chose both the Golden Fleece to symbolize rightful providential rewards for chivalric valor and the pillars of Hercules to signal the superiority of his age over that of the ancients.[1]

Consciousness in the Iberian Peninsula of the preeminence of the moderns over the ancients had more than one manifestation. Iberian humanists were less easily dazzled than their Italian peers by Latin and Greek, and they purposefully set out to develop their own vernaculars.[2] The literatures on metallurgy, medicine, agriculture, surgery, meteorology, cosmography, cartography, navigation, and fortifications studied by Maravall are peppered with comments both on the ignorance of the ancients and on the technical superiority of the moderns. Intellectuals relished every opportunity to remind their readers about all the novelties that had completely eluded the ancients: new empirical breakthroughs in metallurgy (which allowed Spain

PLVS VLTRA

RECIMIENTO DE NAVEGACION
MANDO HASER EL REI NVES
TRO SEÑOR
POR ORDEN DE SV CONSEIO
REAL DE LAS INDIAS
A ANDRES GARCIA DE CES
PEDES SV COSMOGRAFO MAIOR
siendo Presidente e nel dicho
consejo el conde de Lemos

PHILIPPO
TERTIO
HISPANIAR,
REGI

Oceanum reserans nauis Victoria totum
Hispanum imperium claudit utroq̃ polo

A C EIVSDEM
SVPREMO
INDIARVM
SENATVI

A G D   C DD

Fig. 2.3. Frontispiece to Andrés García de Céspedes's *Regimiento de
navegación* (Madrid, 1606). Courtesy of the James Ford Bell Library,
University of Minnesota.

to develop economies of scale in silver mining in the New World); new plants, diseases, and cures; new forms of military maneuver and fortification; and new agricultural techniques.[3] Luis de Camões (1524–80) and Alonso de Ercilla (1533–94) wrote epic poems recording the extraordinary voyages and adventures of real-life Iberian Argonauts and confidently believed that they were superseding Homer. Arguing that Vasco da Gama's deeds dwarfed those of Ulysses, Camões asserted, without inspiring astonishment: "Cesse tudo o que a Musa antiga canta. Que outro valor mais alto se alevanta" (May the ancient Muse be silenced. For greater heroes have now arisen).[4] David A. Lupher has shown that classical Rome became the standard against which Iberians measured their military prowess and scholarly feats in the New World. Conquistadors repeatedly found Julius Caesar and Augustus wanting: petty, insignificant conquerors in a small corner of the world. Poets ridiculed Aeneas as a little hero whose adventures and ordeals had been put to shame by the new Iberian Argonauts. And naturalists dismissed Pliny as a provincial cataloguer of curiosities.[5] It was in this environment critical of the achievements of the ancients that the royal cosmographer Andrés García de Céspedes (d. 1611) published his *Regimiento de navegación* in 1606 (fig. 2.3), the frontispiece to which resembles and foreshadows that later used by Francis Bacon in the *Instauratio magna*.[6]

It is very likely that Bacon (1561–1626) purposefully sought to imitate García de Céspedes, for throughout the sixteenth century, English authors followed the military and technical accomplishments of the Iberians with a mixture of interest and envy. The English acknowledged the technical superiority of the Portuguese and Spaniards when it came to navigation and avidly translated treatises published by Iberian royal cosmographers introducing local audiences to tables and calculations on how to locate latitude (and even longitude) in the open sea. The English sought to imitate the schools for pilots institutionalized in Seville and admired the role of Spanish mathematicians, metallurgists, cosmographers, astronomers, navigators, and hydrographers in the development of empire. As the jealous Richard Hakluyt (1552?–1616) put it in his *Principal Navigations of the English Nation* (1599; smaller first ed. 1589), the Spaniards were far ahead of the English in the colonization of the New World, not only because the former enjoyed "those bright lampes of learning (I mean the most ancient and best Philosophers, Historiographers and Geographers) to showe them light," but also because they possessed "the loadstarre of experience (to wit, those great exploits and voyages layed up in store and recorded) whereby to shape their course." Hakluyt presented the writings of such Spanish cosmographers as Alonso de

Chaves (d. 1587), Jerónimo de Chaves (1523–74), and Rodrigo Zamorano (b. ca. 1545) as exemplary.[7]

It is therefore not preposterous to think that Bacon might have had the Spanish empire in mind when he wrote his *New Atlantis*. The culmination of a lifetime engagement with devising new epistemologies, Bacon's *New Atlantis* describes a utopian island organized around the utilitarian, pragmatic, and experimental manipulation of natural resources. Stranded European sailors happen upon an unknown Pacific island on which the local nobility have instituted a crusading order, Solomon's House. After days of idle waiting on the island, the sailors finally get to meet a powerful member of the order, who introduces them to the secrets of the islanders, namely, the carefully planned, large-scale, mechanical exploitation and reproduction of all natural resources and phenomena. That the islanders and the great lord of the House speak Spanish and that the island is located off the coast of Peru are not the only inklings that Bacon may have had Spanish institutions of knowledge in mind. As Antonio Barrera has elegantly shown, in the sixteenth century, the Spaniards created an extensive culture of empirical, experimental, and utilitarian knowledge gathering that took its cues, not from the classics or the learned, but rather from merchants, enterprising settlers, and bureaucrats.[8]

Settlers and merchants were always on the lookout for new natural resources to sell, constantly hyping the economic windfalls that would accrue to those capable of exploiting the various new mineral, pharmaceutical, and agricultural resources found in the Indies. They also sought to introduce new mechanical devices, demanding patents and monopolies. The Crown responded eagerly but cautiously to all these claims by farming out the testing of the new products and devices to experts back home: physicians, pilots, cosmographers, apothecaries, and inventors. By the early sixteenth century, the scale of claims and counterclaims was such that new institutions had to be created, including the Casa de Contratación in Seville, a veritable "Chamber of Knowledge." Training ships' pilots and assembling credible maps was one of the functions of this new institution; another was apportioning credit among contradictory reports. The Crown also standardized questionnaires and launched large-scale campaigns of data gathering. Finally, the mechanical transformations of landscape undertaken by engineers in the pay of the Spanish empire in Potosí and the central valley of Mexico were as extraordinary as those dreamed up by the knights of Solomon's House, including the creation of artificial lakes and rivers to power mills to crush silver ore and the cutting of sluices through massive

FIG. 2.4. Frontispiece
of *Ordenações manuelinas*
(Lisbon, 1521)

mountains to drain Mexico City (on this, see below). Just like Hakluyt,
Bacon seems to have been well informed about the new Spanish institutions
and practices.[9]

There is also another reason to suspect that Bacon had the Iberians
in mind when he wrote the *New Atlantis*. Like the Portuguese and the
Spaniards, Bacon linked knowledge to crusading. A quick glance at the
sixteenth-century Iberian treatises of cosmography so admired by the English
shows that the Iberians saw knowledge gathering as an expansion of crusading virtues. In his much acclaimed and widely influential *Arte de navegar*
(1545), the royal cosmographer Pedro de Medina (1493?–1567?) asserted
that ships' pilots were the new knights, whose horses were their vessels, and

FIG. 2.5. Frontispiece of Bernardo de Vargas Machuca's *Milicia y descripción de las Indias* (Madrid, 1599). Courtesy of the John Carter Brown Library, Brown University, Providence, R.I.

whose swords and shields were their compasses, charts, cross-staffs, and astrolabes.[10] Astrolabes and coats of arms appear in the *Ordenações manuelinas* of 1521 (fig. 2.4) and other classics of the Portuguese expansion to Africa and Asia—in fact, the armillary sphere and the cross of Saint George are the chief elements of Portugal's coat of arms—and the frontispiece of Bernardo de Vargas Machuca's 1599 *Milicia y descripción de las Indias* (fig 2.5) displays the motto "A la Espada y el compás/Más, y más y más y más" (To the Sword and the compass/More, and more, and more, and more [imperial territorial expansion]). The institutions and values of Bacon's New Atlantis, with its crusading order of Solomon's House, in every respect resemble those created by Spain and Portugal to gather knowledge for utilitarian purposes.

But it is best not to speculate. Juan Pimentel has persuasively shown that Bacon did take key ideas of his *New Atlantis* from the 1606 memorial of Pedro Fernández Quirós's discovery of what he named "Terra Australis." Fernández Quirós (1562–1615) had begun his career in Peru, navigating the Pacific as a pilot for Alvaro de Mendaña, who after having "discovered" the Solomon Islands (1568–69) continued to search for Solomon's Ophir for years. After sailing from Peru in search of Ophir, Fernández Quirós happened upon an island in the Pacific that he took to be a continent. Perhaps thinking to honor the Austrian Habsburg dynastic roots of his reluctant patron Philip III, Fernández Quirós christened the new territory "La Australia del Espíritu Santo." Upon arrival there, he proclaimed the crusading order of the Holy Spirit and set about founding the city of New Jerusalem. Informed by the ideas of the Franciscan Joachim de Fiore, he associated this momentous discovery with the imminent arrival of the new millennium. More important, he envisioned the island as a future utopia where, unencumbered by the learning of the ancients, clear-eyed settlers would collect natural histories. The millenarian, crusading, and utopian empiricist dimensions of Bacon's project are all present in Fernández Quirós's writings. The memorial was translated into several languages, including Latin and English (1617), and Bacon most likely owned a copy.[11]

The *Nova reperta* of the Dutch engraver and painter Jan van der Straet (1523–1605), who latinized his name as Johannes Stradanus, indicates that awareness of the role of Iberians in a dawning modernity was not limited to jealous English imperial competitors.[12] Since his career unfolded in Italy and his patrons were Roman, Florentine, and Genoese clerics and merchants, Stradanus highlighted the contributions to the discovery of America and to the development of the compass of Christopher Columbus (1451–1606), Amerigo Vespucci (1451–1512), and "Flavius Amalfi" (most likely a fictional character, who appears in Italian histories of the compass in the mid sixteenth century) respectively, in the first engraving in his *Nova reperta* (see fig. 2.2). Yet Stradanus's catalogue of nine remarkable new discoveries includes at least three that any of his contemporaries would immediately have granted to the Iberians: namely, the sighting of new constellations in the Southern Hemisphere (the Southern Cross),[13] the introduction of new remedies (the bark of the guaiacum tree), and the discovery of new lands (America). There are also other elements in this picture that contemporaries would not have separated from the history of the Portuguese and Spanish expansions, such as the development of new cash crops and of new military technologies (e.g., the cannon).

Stradanus's *Nova reperta* and Bacon's *Instauratio magna* and *New Atlantis* demonstrate the importance of Iberia to any narrative of the Scientific Revolution. Yet this term has strangely become synonymous with developments in northwestern Europe and the North Atlantic region. When the first histories of the Scientific Revolution by A. Rupert Hall (1954), Marie Boas (1962), and Richard Westfall (1971) appeared in Britain and the United States, they found no room to accommodate Iberia at all. This absence, which is at the core of most metanarratives of the genesis of modernity, derives from the secretive policy of Spain's Habsburg regime, which, as Richard Kagan shows, had a culture of *arcana imperii*, a tendency to keep details of the empire (maps, natural history reports, etc.) unpublished. The secrets leaked, but the massive collected learning related to empire building was destined to circulate only in manuscript form.[14] Even the Bourbons, who set out to do away with many such Habsburg practices in the eighteenth century, failed to publish the countless studies of cartography and natural history they sponsored.

In addition to being marginalized by the practice of privileging scribal over print culture in order to keep knowledge secret, the science sponsored by the Spanish-Portuguese state was of a kind historians of science like Hall, Boas, and Westfall considered peripheral to the narrative of the Scientific Revolution. For these historians, astronomy, physics, and mathematics held the key to explaining the transformations of early modern understandings of nature. The mathematization and mechanization of the cosmos in the seventeenth century ultimately led to secularism, industrialization, and capitalism: the birth of the modern world. The sciences the Iberian powers sponsored, mostly cartography and natural history, were thought not to have been related to these economic, religious, and cultural changes. Yet there is today a growing realization that natural history and cartography were not entirely peripheral to the momentous epistemic transformations of modernity. Some historians of science have therefore turned to study the sciences sponsored by Iberian empires.[15] But interest in Spain and Portugal is still marginal. Why?

The reason lies, ultimately, in the narratives of modernity inaugurated first by Protestantism and later by the Enlightenment, both profoundly hostile to Catholic Iberia. Some one hundred and fifty years after Bacon borrowed the tropes and motifs to depict the arrival of modernity from García de Céspedes, Zacharie de Pazzi de Bonneville (ca. 1710–ca. 1780) has his "philosophe La Douceur" declare that "there is no nation more brutish [*abruti*], ignorant, savage, and barbarous than Spain."[16] In 1777,

Joseph La Porte (1713–79) concluded in his *Le voyageur françois* (1766–95), a massive compilation of travel narratives presented as letters from a fictional roving observer to a lady back home, that Spain was a land of superstitious folk, still practicing sciences inherited from the Moors, namely, "judicial astrology, Cabala, and other Arab inanities." Spaniards, he further argued, had boundless admiration for Aristotle, "whose senseless and tenebrous philosophy" they blindly followed.[17] Finally, in 1781, the abbé Raynal (1713–96) maintained in his *Histoire philosophique et politique des établissements et du commerce des Européens dans les deux Indes* that "never has a nation been as enslaved to its prejudices as Spain. In no other place has irrationality [*le déraison*] proven as dogmatic, as close, and as subtle."[18] Ever since the eighteenth century, Iberians have come to represent the antithesis of modernity.

The erosion of the Iberian empires in the face of increasing Dutch and English competition and the failure of Spain and Portugal to carry out reforms to consolidate the centralizing power of the state, as in France, led to the relative decline of the Iberians in the seventeenth century. Already during the Reformation and wars of Dutch independence, northwestern European printers had created an image of the Iberians as superstitious and rapacious plunderers. The decline of the Iberian empires not only hardened this perception; criticism now came wrapped in the language of progress, and the Iberians were cast as essentially non-Europeans: backward and ignorant.[19] By 1721, Montesquieu (1689–1755) could maintain in the pages of *Lettres persanes*, without being challenged, that Spaniards were only good at writing chivalric romances and dogmatic scholastic treatises.[20] Not surprisingly, none of the metanarratives of modernity and progress that came of age in the eighteenth century have found a place for the technological and philosophical contributions of the Iberians in the early modern period.

This tendency to neglect the contributions of the Spanish to natural philosophy in the sixteenth century can be observed in Frances López-Morillas's translation into English of José de Acosta's *Historia natural y moral de las Indias*, published by Duke University Press (2002). For all its merits, the new edition seems more preoccupied with Acosta the historian and anthropologist than with Acosta the natural philosopher.

"And my desire is that all I have written may serve to make known which of his treasures God Our Lord divided and deposited in those realms; may the peoples there be all the more aided and favored by the people of Spain, to whose charge divine and lofty Providence has entrusted them." With these words in the dedication to Philip II (1527–98), José de Acosta (1540–1600) summed up the spirit of his *Historia*, for the celebrated Jesuit was interested both in explaining the conquest as a preordained providential event and in identifying signs of intelligent design in the many natural wonders of the American continent. Acosta was a man of omnivorous curiosity, with an uncanny ability to find divine order in contingency, chaos, and probability. But he was not simply a Christian philosopher. As the above quotation makes clear, he was also a pragmatist interested in how things work and how colonial peoples thought, so as to use and manipulate the former and to convert and govern the latter.

Acosta's views have been available to English-speaking audiences for centuries. Unlike the writings of scores of other sixteenth-century Spanish, Creole, mestizo, and Amerindian authors whose treatises on the natural wonders of the Indies and the past of local indigenous peoples commanded little attention until recently, Acosta's *History* was immediately translated into several European languages, including English. In the eighteenth-century Atlantic world, when all sources produced in the early Spanish empire came to be seen as untrustworthy and useless, only Acosta's treatise was deemed worth reading. Despite Acosta's reputation, however, since 1604, the average English-speaking reader interested in the Jesuit has had to plow through Edward Grimeston's translation. Grimeston's prose in *The natvrall and morall historie of the West Indies* may have served Elizabethan audiences well, but today it seems stodgy and distant. Fortunately, students can now turn to López-Morillas's crisp new rendition.

López-Morillas's translation has been edited by the Andeanist Jane E. Mangan and is accompanied by an introduction and commentary by the literary critic Walter D. Mignolo. Both Mangan's annotations and Mignolo's study help put Acosta in a larger cultural and ideological context, but their approach betrays a bias that is typical of most contemporary scholarship in the field.

In 1604, Grimeston and his Tudor and Stuart audiences considered Acosta to be not only a keen observer of things Amerindian but a great natural philosopher. Yet by the late seventeenth century, Acosta began to be read only for his contributions to anthropology and ethnography. Students and scholars today do not turn to Acosta for information about the nature of the stars and heavens in the Americas but to reconstruct the lives of Amerindian peoples and the nature of colonial power. What is left out, however, are the questions that most captivated Acosta: why tides and winds in the Southern and Northern hemispheres have different timings and directions; why the torrid zone of Peru enjoys a temperate climate year round, instead of scorching heat; why seasons of rain and drought follow exactly opposite patterns in Europe and Peru; why mercury attracts silver; and so on. Three out of five pages in Acosta's *Historia* are devoted to accounting for the seemingly puzzling behavior of the cosmos in the Indies. Acosta sets out to prove that nature in America, just as much as in Eurasia, is a docile servant of God, following predictable laws. For all their contributions, Mangan and Mignolo deal only tangentially with this essential facet of Acosta's world.

The disregard for Acosta as a natural philosopher and for Spanish science at large is also obvious in Lorraine Daston and Katherine Park's otherwise marvelous book *Wonders and the Order of Nature, 1150–1750*. Although they make passing and superficial references to Nicolás Monardes (ca. 1512–88) and one Ferrando de Oviedo (actually Gonzalo Fernández de Oviedo [1478–1557]), Daston and Park completely overlook the Spanish literature on the New World's wonders, including, for example, Juan de Cárdenas's path-breaking *Problemas y secretos maravillosos de las Indias*, published in Mexico in 1591, Baltazar Dorantes de Carranza's *Sumaria relación de las cosas de Nueva España* ([1604] 1902), the massive seventeenth-century natural history of marvels by the Jesuit Juan Eusebio Nieremberg, *Historia natvrae, maxime peregrinae* (1635), the capstone of this tradition, and Antonio de León Pinelo's *El paraíso en el Nuevo Mundo* ([1645–50] 1943).[21] Daston and Parker show absolutely no awareness that Spanish natural histories of the Indies like Acosta's were attempts at modifying dominant narratives of marvels. A firm believer that demons were particularly powerful in the Indies and largely responsible for the idolatry of the colonized natives, Acosta was nevertheless not willing to cede the realm of the natural and the marvelous in the New World to the devil.[22] His *Historia natural y moral* constantly seeks to frame puzzling natural phenomena and the seeming inversion of physical laws in the Indies with a discourse of providential design and lawful regularities. By so doing, Acosta sought to steer early modern European perceptions of nature in the New World away from the realm of the preternatural and thus the demonic.[23] This is all the more remarkable because Acosta was a firm believer in the imminence of the Apocalypse, a fact that goes completely unnoticed by Mangan and Mignolo. Acosta's *De temporibus novissimis libri quatuor*, originally published in Rome in 1590, sought to prove, among other things, that the multiplying instances of witchcraft, demonic possessions, prodigies, and the preternatural in early modern Europe were all signs of the devil having been unleashed on the eve of the Apocalypse, as suggested in the Bible.[24]

Examples of how the history of Renaissance Iberian natural philosophy is usually overlooked abound. Given the dominant narrative of the North Atlantic origin of modernity sketched above, this is not at all surprising. It is surprising, however, that there is little room in the vibrant new field of the history of botany, natural history, and empire for considering how Portugal and Spain set long-term European patterns. Take, for example, Richard Drayton's account of the rise of the modern botanical garden in his *Nature's Government: Science, Imperial Britain, and the "Improvement" of the World* (2000). Drayton offers a dazzling, brilliant account of how botany and empire developed in tight, mutual interaction during England's eighteenth and nineteenth centuries. Although Drayton's focus is on the history of the Royal Botanic Gardens at Kew, he spends considerable time exploring the origins of the early modern botanical garden, which he traces to particular intellectual, religious, and political forces. Intellectually and religiously, the culture of the botanical garden was tied to humanist efforts to catalogue the world and recover Adam's long-lost grip on creation.[25] Politically, the garden first sought to glorify monarchs as new Solomons—learned kings deeply concerned with the secrets of nature so as to benefit the local commonwealth. Later, these monarchical philosophical pretensions gave way to a culture of ornamentation, in which all sorts of powerful patrons, not only rulers, set out to collect plants and tend gardens to dazzle, while consolidating power and prestige. In these various phases, Drayton reminds us, naturalists, monarchs, and humanists found the plants they needed overseas. But this very process of primitive accumulation of botanical knowledge changed ideas about the polity, religion, and the order of nature. Eventually, the botanical garden became an institution that helped generate colonialist ideologies, while promoting large, global agricultural economies of scale.

This genealogy of the botanical garden (from medicine to ornamentation to plantation agriculture) does not pay sufficient attention to developments in sixteenth-century Portugal and Spain, however, for early modern botany was as rooted in the entrepreneurial, utilitarian efforts of apothecaries in gardens and hospitals in Seville, Lisbon, Goa, and New Spain as in the humanist culture of the medical faculties of Padua, Leiden, and Montpellier. From the time of its revival as a science in the Renaissance, botany served the needs of transnational merchant capital. Take, for example, the case of Carolus Clusius (1526–1609), whom Drayton refers to as "the Copernicus who shattered the Hellenocentricsm of Renaissance botany."[26] The dean of early modern botany and the founder of the botanical gardens of Vienna and Leiden, Clusius spent his life chasing after exotica to expand the classical repertoire of European botanical knowledge, limited for centuries to the few hundred plants catalogued in the works of Theophrastus and Dioscorides. As Drayton himself correctly points out, Clusius made available in Latin the works of Portuguese and Spanish doctors and apothecaries such as Nicolás Monardes, García d'Orta, and Cristobal Acosta (ca. 1515–ca. 1592). These, however, were all works single-mindedly focused on the potential commercial value of newly discovered plants, with little use for speculative philosophy.[27] Clusius inherited a keen eye for the utilitarian, commercial value of exotic commodities from these Portuguese and Spanish treatises, as well as from his travels through the various botanical gardens of Portugal and Spain.[28] Thus Clusius's additional notes to Orta's short treatise on Southeast Asian aromatic and pharmaceutical plants also included references to potentially profitable exotica collected by Francis Drake (1540?–96) in his recent circumnavigation of the globe.[29] Clusius's translation into Latin of Thomas Harriot's description of Virginia for Theodore de Bry (1528–98) is not a simple rendition. It includes long lists of goods used by the local inhabitants, as well as of botanical staples with potential commercial value.[30]

The utilitarian, pragmatic, commercial aspect of Iberian natural history found its culmination in the expedition and work of Francisco Hernández. Sent by Philip II to gather material for a natural history of herbals, Hernández spent seven years in Mexico (1571–77) experimenting in hospitals for natives established by the Spanish clergy and interviewing Nahua intellectuals versed in Latin. By the time of his return, he had assembled eleven volumes of illustrations of 3,000 different species of plants (as well as of minerals, animals, and local antiquities) and several other volumes of text. Phillip II felt the work to be too philosophical and asked his royal physician, the Neapolitan Nardo Antonio Recchi, to plow through Hernández's work to come up with a list of useful pharmaceutical plants. Philip died while Recchi was at work on this, however, and neither Recchi's anthology nor Hernández's massive natural history was ever made available by the Crown, which at the time was seeking to shut down all publications on the Indies in an effort to deny rival powers any additional knowledge of, and footholds in, the New World. Recchi's compilation had to wait some sixty years to appear, this time with various appendices and notes by several members of the Accademia dei Lincei.

The history of the Academy of the Lynx and of the publication of Recchi's manuscript is told by David Freedberg in *The Eye of the Lynx*, a most learned and lavishly illustrated book (2002). In Freedberg's narrative, this Roman academy, to which Galileo belonged, appears, again, as *the* harbinger of modernity. Mostly composed of German, Roman, and Neapolitan scholars, led by Federico Cessi (1585–1630), the academy edited works that sought to find the hidden order of nature. With the aid of microscopes, telescopes, and the art of dissection, academicians like Galileo and Cessi observed not only the surface appearance of plants, fossils, insects, and stars but also their internal, intimate structure. In so doing, these *novatores* set out to undermine the authority of the ancients and radically altered the way knowledge was gathered and classified. As part of this new effort, the academicians were particularly interested in curiosities and exotica, and when they learned that Recchi's nephew, one Marco Antonio Petilio, had kept the doctor's manuscripts, they pounced on them. Over the course of some forty years, at different times, Johannes [Terrentius] Schreck (1576–1630), Johannes Faber (1570–1640), Fabio Colonna (1567–1650), Francesco Stelluti (1577–1653), and Federico Cessi continued to add notes and marginalia to the original manuscript, until it finally came to light in 1651 under the title of *Rerum medicarum Novae Hispaniae thesaurus.*

A quick glance at the frontispiece (fig. 2.6) demonstrates the utilitarian emphasis of the whole enterprise: scantily clad natives offer Philip IV (1605–65) (represented by his coat of arms) the botanical riches of Mexico, which (pace Drayton) are simultaneously medicinal, ornamental, and agricultural. Viewers are invited to step through the doorway and into one of the territories of the Spanish empire, which also include Castile, Leon, Granada, Portugal, Sicily, Naples, Aragon, Flanders, Jerusalem (!), Mexico, and Peru (see the coat of arms in the frontispiece in *Rerum*, originally designed in 1628, when Portugal was still part of the Spanish empire; the serpents and eagles below the Portuguese coat of arms would seem to represent Mexico and Peru). It is useful at this stage to remind the reader that Cessi and his Roman and Neapolitan allies were subjects of the loose Spanish monarchy.[31] The utilitarian, commercial emphasis of the work surfaces repeatedly in the text itself. The printer Giacomo Mastardi, for example, ends his preface to the reader, in which he outlines the complex, tortuous history of the manuscript and identifies the various contributors, by insisting that "not only the herbalist, the lover of natural history, the medical doctor, the philologist, the taxonomist [*Phytosophus*], and the collector of ornamental flowers for princes and noblewomen . . . but also the shopkeeper [*institor*], the quack [*pharmacopola*], the apothecary [*pharmacis mercator*], and the perfumer [*odorarius*], whether in search of health, pleasure, or money, will find a wealth of objects, images, and names to satisfy your mind, eyes, and desire."[32] Johannes Schreck (aka Johannes Terrentius), charged with adding glosses and commentary to Recchi's original manuscript, defends Recchi's decision to follow Theophrastus and Dioscorides on pragmatic, commercial grounds. Thus in the section on Mexican trees, Terrentius argues that dividing plants into trees, bushes, and herbs is justifiable in the case of trees, because trees are the source of many riches, give us shelter from the attack of beasts, keep us from drowning in tempests, and allow us to ply the menacing seas to discover new lands and ultimately engage in commerce.[33]

For all his enormous contribution, Freedberg excludes Iberia from his history of the Accademia dei Lincei and particularly from the history of the publication of Recchi's manuscript. In Freedberg's narrative, Spain appears as an obstacle, and he presents Hernández as incompetent, barely capable of organizing the material he collected (247). Ricchi appears as a physician cowed into silence by the Spanish king, fearful of sharing the fruits of his labor with others (249). Although he acknowledges that it was a Spanish official in Rome, Alfonso de Las Torres, who finally put up the money to publish the work, Freedberg does not make much of this, nor of the fact that Rome and Naples were at the time cultural satellites of Spain.[34] Yet the very evidence Freedberg presents shows that Spain was a willing participant throughout the many years it took the Accademia dei Lincei to bring the book to light. Linceans like Johannes Heckius (1576–ca. 1618) (253) and Cassiano dal Pozzo (1588–1657) (262) had repeatedly had access to Hernández's manuscripts and illustrations at El Escorial. Moreover, when Cassiano visited Madrid in 1626, he obtained the Codex Badianus, an illustrated herbal written in Latin in 1552 by two Nahua intellectuals, Juan Badiano and Martín de la Cruz, as a present for the pope.[35] More important, Freedberg unwittingly shows that the academicians received help at every turn from learned Spanish or Creole clerics living in Rome, who again and again provided Linceans with animals, plants, documents, and much-needed interpretations (261, 265).[36] Freedberg's work demonstrates how difficult it has become for Anglo-American scholarship to bring Iberia back into narratives on the origins of modernity.

Spaniards, to be sure, have not taken kindly to this neglect. Spanish intellectuals compiled massive bio-bibliographies to demonstrate the remarkable intellectual successes of the Spaniards since the Romans. Patriotic compilations such as the *Bibliotheca Hispana vetus* (1696) and *Bibliotheca Hispana nova* (1672) of Nicolás Antonio (1617–84) multiplied in the eighteenth and nineteenth centuries, particularly those on sixteenth- and early seventeenth-century authors and texts. With the support of Dutch and German printers and scholars, a Valencian scholar, Gregorio Mayans y Siscar (1699–1781), catalogued the achievements of early modern Spanish arts and science,

FIG. 2.6. Frontispiece to Francisco Hernández's *Rerum medicarum Novae Hispaniae thesaurus* (Rome, 1651). Reproduced by permission of the Huntington Library.

inventing a new period in Spanish history, the *siglo de oro*, or "Golden Age."[37] These writings, unfortunately, led nowhere. They were consumed within Spain and deployed to bolster various patriotic agendas. The Spanish Golden Age today evokes the names of Diego Velásquez (1599–1660) and Lope de Vega (1562–1635). We associate early modern Spain with painters and poets, not metallurgists and astronomers.

José María López Piñero's *Ciencia y técnica en la sociedad española de los siglos XVI y XVII* (1979) has perhaps been the most significant and influential study of the history of early modern Spanish science and technology to appear in the last quarter of the twentieth century. It is a formidable effort, wide-ranging and ambitious in scope. The book is unevenly divided into four sections. The first is slim and devoted to issues of historiography and methodology. It surveys the literature that since Nicolás Antonio has sought to reconstruct the accomplishments of sixteenth-century science in Spain. Most of these studies, López Piñero maintains, were patriotically biased. Many were erudite but narrowly focused on one or two disciplines. López Piñero presents his book as the first comprehensive, unbiased study, entirely based on meticulous research from primary sources in all disciplines. Section two is a statistical and a historicist study of practitioners of science and technology in sixteenth-century Spain. It defines science according to early modern criteria and shies away from anachronistic categories. It identifies the institutions in which science was done in Spain and seeks to quantify this work according to such variables as the number of published texts per field and the regional origin of authors and publications. This section also contains studies of the social standing of the various practitioners according to trade, profession, and ethnicity (Christian, Jewish, converso, and Morisco).[38] Finally, by looking at patterns of self-imposed cultural isolation on the part of universities and the Crown, encouraged by the Inquisition, it offers suggestions as to why Spanish science and technology had begun its inexorable decline by the late sixteenth century. The third section, the bulk of the book, is a veritable encyclopedia, a painstaking discussion of hundreds of texts and authors according to disciplines: mathematics, cosmography, navigational science, geography, natural philosophy, engineering, metallurgy, natural history, anatomy, medicine, and surgery. Section four, much slimmer, jumps to the late seventeenth century to study innovators in places like Valencia, Madrid, and Seville who sowed the seeds of eighteenth-century scientific renewal by embracing the empirical-experimental practices and philosophical categories of the new mechanical philosophy. This section, also encyclopedic in scope, is organized around authors, not disciplines, namely, the *novatores* (innovators)

Juan Bautista Juanini (1636–91), Crisóstomo Martínez (ca. 1638–ca. 1694), Juan de Cabriada (1660–1730), Joan d'Alós, Jaime Salvador, Juan Caramuel Lobkowitz (1606–82), Vicente Mut (1614–87), José de Zaragoza (1627–78), Tomás Vicente Tosca (1651–1723), Juan Bautista Corachán (1661–1741), and Antonio Hugo de Omerique.

López Piñero's study of sixteenth-century Spanish silver-mining technologies typifies his approach. In an impressive command of both primary and secondary literature, López Piñero identifies a series of errors that have plagued the literature on the subject ever since the publication in 1786 of the first history of technologies of amalgamation by the Austrian mineralogist Ignaz von Born (1742–91), *Über das Anquicken der old- und silbehaltigen Erze, Rohsteine, Schwarzkupfer und Hüttenspeise* (Of the Amalgamation of Gold and Silver Ores [etc.]). Beginning with Born, it has been common to argue that the process of extracting silver from silver ore with mercury involved a single "discovery." There is some debate as to where that "discovery" took place; some have argued that it was a well-established ancient practice; others that it was originally tried by German miners, who took it with them first to Spain and later to the New World; and still others, particularly Spaniards and Spanish Americans, that it was invented in the New World (with some debate as to whether this was in Peru or Mexico). López Piñero shows that the process of amalgamation was not reproducible from place to place and that therefore it was a complex practice that had to be "discovered" time and again. Although mixing mercury and silver was a well-known alchemical procedure, extraction of silver *from silver ore* was a new technology first developed by Spaniards, not Germans. This technology, however, required persistent experimentation, because the amount of mercury needed varied according to the nature of the local ore. It was also an industrial process; it needed to be standardized and reproduced in economies of scale. Bartolomé de Medina (1492–1585) achieved the first successful trial in Pachuca, Mexico, in 1555. Yet his procedures and formulas (they included crushing the ore and mixing it with salt, copper/iron sulfides, and mercury) did not travel well. They did not work in Spain (1555–62) or Germany (1588), and between 1559 and 1568, they did not work in Peru either. It was only in the 1570s that the miner Pedro Fernández de Velasco, through controlled experiments with local ore, discovered successful new formulas and techniques, triggering a silver rush in the Andean highlands that in the space of a few years created the largest city in the Spanish empire, Potosí. López Piñero closes his brief study of amalgamation describing the new technologies introduced by Juan Capellín (1576) in Tasco, Mexico (which shortened the time the amalgams "rested" in *patios* from twenty to four days); by Carlos Corzo (1587) in Potosí (which reduced the amount of mercury needed by introducing new iron sulfites into the original reaction with salt and crushed ore);[39] and by Alvaro Alonso Barba (ca. 1600) in Potosí (which improved the amount of silver recovered from the amalgam by heating the combination of sulfites, mercury, and crushed ore).

In fewer than six pages (259–65), López Piñero takes on well-established scholarly traditions and offers a comprehensive history of amalgamation in the New World.[40] Moreover, he offers tantalizing suggestions for future research. López Piñero insists that all experimentation and innovation in amalgamation took place in the New World, not Europe. There are remarkable similarities between this case on the imperial periphery and the metallurgical experimentation in the ironworks of seventeenth-century New England. William Newman has shown that figures like George Starkey (1627–65) took advantage of the very marginality of New England in the British empire to introduce rigorous experimental procedures (tied to accounting measures in proto-industrial firms) that transformed the content and practices of seventeenth-century alchemy at the European core.[41]

There are some aspects of López Piñero's book that the above example on metallurgy should make obvious. First, the book aspires to be an encyclopedia: López Piñero privileges sorting, weighing (credibility), counting, identifying, and cataloguing over grandiose speculation. Second, the book has no other overarching argument besides insisting on the vitality of Spanish science in the sixteenth century through the accumulation of encyclopedic examples. Finally, Spain is thought of in the wider Atlantic context. The New World occupies center stage.

For all its claims to objectivity and scholarly detachment, López Piñero's book, like the genre of bio-bibliographies exemplified by Nicolás Antonio, is organized around the study of authors and texts (particularly the third and fourth sections) and motivated by patriotism. Like that other Valencian Mayans y Siscar, López Piñero is enthralled by the idea of a sixteenth-century Spanish golden age of erudition and creativity in the sciences and technology. Like Mayans, he finds the hegemonic North Atlantic historiography on the Scientific Revolution by and large unaware of Iberian contributions to it.[42] Aware that his book could be construed as yet another in a long list of patriotic surveys, López Piñero devotes the first section to establishing differences with this literature. But there is much in López Piñero's study that resembles the genre he dismisses. Like the eighteenth-century monumental studies by Juan Francisco Masdeu (1744–1817), Francisco Xavier

Lampillas (1731–1810), and Juan Andrés (1740–1817) on ancient, medieval, and contemporary Spanish literature, done to correct claims of the negative influence of Spanish bad taste in Italian literatures from Martial to Lope de Vega, López Piñero offers his readers a veritable encyclopedia.[43]

The encyclopedic aspirations of López Piñero come through most clearly in his editorial efforts. The Instituto de Historia de la Ciencia y Documentación (Institute of the History of Science and Bibliographic Documentation, now named after López Piñero) at the University of Valencia, has over the past four decades published some sixty volumes in three different series: monographic studies, editions of primary sources, and that most traditional of Spanish genres, bio-bibliographies. In addition to organizing this massive editorial effort, López Piñero has also put together the invaluable two-volume *Diccionario histórico de la ciencia moderna en España* (Historical Dictionary of Early Modern Spanish Science) (1983), in collaboration with Thomas Glick, Víctor Navarro Brotóns, and Eugenio Portela Marco.

Typical of this effort to publish primary sources is Lopez Piñero's edition of the beautiful Codex Pomar (2000). Philip II was a generous patron of natural history. His palace at El Escorial had eleven rooms full of ovens and glassware, large alchemical laboratories devoted to distilling the quintessence of plants. The "philosophical tower" was the jewel of Philip's laboratory: a twenty-foot glass tower capable of producing two hundred pounds of liquid herbal extracts every twenty-four hours. To keep his alchemical laboratories running, Philip established botanical gardens in and near El Escorial. The king also kept pleasure gardens, where in lavish displays of engineering prowess, he manipulated water and earth like Hercules and Orpheus. Lest his commitment to natural history ever be doubted, Philip sent expeditions to the New World and kept birds and quadrupeds, including such exotica as a rhino (which commanded admiration, even though it was hornless and blind), in his palaces. Philip II had at his disposal plants and animals from every corner of the planet, for he ruled over the largest empire ever assembled, with colonies in Africa, India, China, Europe, and the New World. In his sumptuous edition of the Codex Pomar, López Piñero has, once again, made it possible for historians to glimpse the world of medical herbals and natural histories at the court of Philip II.

A collection of 218 hand-colored illustrations of plants (148) and animals (70) from Europe, the New World, and Asia, the Codex Pomar originally belonged to Philip II. Seeking to lure Jaume Honorat Pomar (ca. 1550–1606), one of the leading luminaries of the medical school of the university of

Valencia, to Madrid, Philip II gave Pomar the codex as a gift. After having taught anatomy (inspired by the humanism of Vesalius) for four years, and after having led the chair of herbals for fourteen, Pomar finally relented and left Valencia to serve Philip II in Madrid the very year the king died (1598).

In the introduction to this edition, López Piñero goes over material he has already covered in other publications: the tradition of alchemical-herbal medicine in Spain; the humanism of the Valencian medical school; and the contributions to natural history of Nicolás Monardes and Francisco Hernández (on which see further below). Surprisingly, there is precious little here on the codex itself. Although López Piñero, with the erudition that has always characterized his work, locates each of the animals and plants in the illustrated codex within ancient and contemporary (Dioscorides, Pliny, Covarrubias) and Linnaean modern taxonomies, readers are largely left on their own to interpret the 218 images. What do the images tell us about the culture of patronage, diplomacy, gardens, and menageries in late sixteenth-century Spain? There are, for example, six gorgeous images of tulips. Were these collections of tulips ever used as a diplomatic tool to deal with the Ottomans? A beautiful illustration of a rhino's horn parades before our eyes. Where did the rhino come from and how did it contribute to enhancing the prestige and ritual power of the king? The introduction also does not help us to understand how the codex fits into sixteenth-century European traditions of plant illustrations. Most of the plants in the codex are shown at different stages in their natural life cycles; some seem to represent pressed plants, not live specimens. Were these typical illustrating techniques?

Some twenty-three years after its publication, *Ciencia y técnica* remains highly influential within Spain. The methodology, insights, and encyclopedic sensibilities that López Piñero first deployed in his treatise surface now in almost every study on any subject in the field of early modern science and technology. Consider, for example, the superb *Técnica y poder en Castilla durante los siglos XVI y XVII* (1989) by Nicolás García Tapia, in which the author seeks to reconstruct a long-forgotten chapter in the history of sixteenth-century northern Castilian engineering of roads, bridges, canals, aqueducts, watermills, looms, ironworks, public buildings, and water pumps. Like López Piñero's, his approach is comprehensive and erudite. García Tapia ransacks the archives of Simancas, Valladolid, Burgos, Medina del Campo, Palencia, Salamanca, and Segovia for evidence. The results confirm many of López Piñero's findings, including that sixteenth-century Spain witnessed much innovation, only to experience a decline in the following century; that northern Castile was part of larger European networks, exporting hydraulic engineers to the rest of the continent and relying on imports or industrial spying to develop other technologies; and that a number of striking technical innovations did not first take place in Italy or northwestern Europe. García Tapia locates several patents by northern Castilian inventors of steam engines, diving gear, submarines, and water turbines, for example. Finally, like López Piñero, García Tapia shies away from sweeping arguments or large interpretations; his goals are modest; he simply seeks to add one brick to the edifice of knowledge.

This attention to erudition also permeates the work of López Piñero's colleague Víctor Navarro Brotóns, who has for several decades contributed mightily to the editorial and encyclopedic efforts of the Institute of Bibliographic and Historical Studies on Science at the University of Valencia. A leading authority on the history of early modern Spanish astronomy and mathematics, Navarro Brotóns has studied the reception of Nicolaus Copernicus (1473–1543) in sixteenth-century Spain. Typical of his approach and sensibility are his publications on the "Copernicans" Diego de Zúñiga (1536–1600) and Jerónimo Muñoz (d. 1591).[44]

Zúñiga, a polymath biblical scholar and professor of Holy Scripture at the University of Osuna, is known among specialists on Copernicus for having embraced the physical reality of heliocentrism as early as 1584 in an interpretation of Job 9:6 ("God moves the Earth from its place and makes its columns tremble") entitled *Job commentaria*. In fact, Zúñiga was not alone: Copernicus's *De revolutionibus* (1543) made it onto the reading lists at the University of Salamanca as early as 1561. These facts throw a monkey wrench into the North Atlantic narratives of modernity. How could a closed, backward, post-Tridentine country like Spain be at the forefront of the Copernican revolution? Fortunately for the survival of these North Atlantic discourses, the Spanish Inquisition clamped down on Zúñiga's work in 1616, although it never banned *De revolutionibus*. Oddly, this story has served to reinforce the very North Atlantic narratives that Zúñiga's alleged intellectual precocity threatened. Along with Galileo's, Zúñiga's name is bandied about to exemplify the retrograde medieval impulses of a dogmatic Catholic Church.

Navarro Brotóns does not believe any of this. In his study, Zúñiga is depicted as a self-aggrandizing biblical scholar who, in order to impress patrons at court and Rome, used Copernicus to offer a dazzlingly novel exegesis of this passage of Job. Having been told that heliocentrism did not conform to contemporary perceptions of physical reality, however, he immediately set out to write one of the first essays in Europe denouncing the theory as scientific nonsense, his *Philosophia prima pars* (1597). This in fact was part of larger pattern in Spain: Copernicus was well received among astronomers, who found his calculations useful. Spanish astronomy and cosmography, one would be right to infer, were particularly pragmatic, because they developed not only within universities but also at court and in special academies to train ships' pilots, in the context of relentless imperial expansion.[45] Whatever worked to calculate the motion of planets and stars was welcomed. But when Copernicus's theory became part of larger philosophical and biblical debates in the 1590s, all Spanish intellectuals, including Zúñiga, quickly opted to reject Copernicus and embrace "common sense": the Aristotelian notion of an Earth at rest.[46]

Navarro Brotóns does not really offer any all-encompassing interpretation of Spanish astronomy. Pragmatism prompted by the needs of imperial expansion might be a useful theory to account for why Copernicus's mathematical model was particularly welcomed in sixteenth-century Spain, yet Navarro Brotóns is not really looking for great explanatory models. He simply identifies the facts in painstaking detail, author by author, text by text. His *Matemáticas, cosmología y humanismo en la España del siglo XVI* (1998) is, in the same vein, a study of the life and works of another controversial Spanish scientific figure, the Valencian astronomer Jerónimo Muñoz. Like Zúñiga, Muñoz was a humanist, so good at languages, particularly Hebrew, that there was some suspicion he was a converso (i.e., a Jew feigning conversion to Christianity to escape the Inquisition). Be that as it may, Muñoz got his training in France, Flanders, and Italy and returned to teach astronomy and mathematics first in Valencia and later in Salamanca. All his life he was part of larger European corresponding networks (which included Tycho Brahe [1546–1601]). His *Libro del nuevo cometa* (1573) made the first forceful case for the recent origin of a star, the supernova of 1572, demonstrating its supralunar origin.[47] His observations led him to challenge Aristotle's thesis of the incorruptibility of the heavens. Like Zúñiga, Muñoz read and praised Copernicus for pragmatic purposes, but in his unpublished "Commentaria Plinii libri secundi" (1568), Muñoz sought to refute the physical plausibility of heliocentrism. Navarro Brotóns traces Muñoz's works in painstaking detail and, after collating two extant manuscripts of the "Commentaria," offers an annotated edition (and a Spanish translation) of it. Like López Piñero and García Tapia, Navarro Brotóns emphasizes getting the facts straight, not daring interpretations.[48]

Although many other studies of Renaissance science and technology by Spanish historians could be included in this review, there is enough here to advance a handful of generalizations.[49] It is clear that López Piñero's *Ciencia y técnica* had a considerable impact in shaping a particular research agenda and epistemological style.[50] In most Spanish studies, a premium is put on erudition and on meticulous reconstruction of past events through painstaking sifting of texts and archival research. Authors shy away from grand interpretations. The sixteenth century is particularly well regarded, because most authors seek to present Spain as European, partaking of larger continental intellectual movements. By the same token, the seventeenth century is something of an embarrassment for most authors, for at this time Spain allegedly withdrew and went into a spiral of intellectual isolation and decline. It is not that Spanish historians of science ignore the seventeenth century. On the contrary, they are always looking for evidence of innovation and revival. Navarro Brotóns, for example, has sought to reconstruct some aspects of the history of physics and mathematics during Spain's age of "decline" by examining the history of the Colegio Imperial (Imperial College), a court institution led by the Jesuits, that most cosmopolitan of religious orders.[51] López-Piñero, as noted, also devoted one entire section of his *Ciencia y técnica* to the science of the seventeenth century. His interests, however, are narrowly limited to identifying those who brought Spain back onto the modern path by embracing the new mechanical philosophy. The issue is that by looking only for traces of modernity in the Spanish polity, these scholars ignore all other aspects of the practice of natural philosophy in early modern Spain. The allegorical and emblematic sciences of the baroque that flourished in the seventeenth century have been forgotten, for example, cast aside as a shameful chapter in Spanish history (on this tradition, see chapters 3 and 4 below).

There is a clear teleological thrust to the scholarship I am reviewing: the disciplines and practices explored are those that can be linked to the genealogy of European modernity. Let me be clear: Spanish scholars are perfectly aware that most modern disciplines looked rather different in the Renaissance; there were no chemists but alchemists, no physicists but astrologers, no biologists but collectors of cabinets of natural history and curiosities. But for all their historicist sensibilities, most Spanish historians of science seem overly concerned with identifying the pioneers of modernity. Moreover, for all their emphasis on placing Spain firmly within wider European traditions, they have little to say on Portugal. This is surprising, given the fact that Renaissance science in Spain was deeply influenced by fifteenth- and sixteenth-century Portuguese traditions in cosmography, astronomy, and navigation.[52] More important, from Philip II to Phillip IV, the empire was as much Portuguese as it was Spanish. But despite their emphasis on exploring the intellectual genealogies of the modern nation, Spanish historians of Renaissance science and technology have long been interested in understanding developments in colonial Spanish America.

López Piñero, again, has been instrumental in cultivating this Atlantic sensibility. He and his Valencian colleagues have made great strides in understanding the process by which botanical and pharmaceutical knowledge collected by Spain in the New World spread to the rest of Europe.[53] Valencians have been particularly interested in reconstructing how the work of the naturalist Francisco Hernández moved around Europe. A typical humanist, steeped in classical learning, who befriended such luminaries as Benito Arias Montano (1527–98) and Francisco Valles (1524–92), Hernández went well beyond Philip II's request for exotic plants for the royal pharmacy and put together a mammoth natural history of New Spain that took seven years to complete (1571–77) and included descriptions of some 3,000 new species of plants (compared to some 350 inventoried by Theophrastus and 500 by Dioscorides).

Scholars had long assumed that Hernández's work reached other Europeans only through Juan Eusebio Nieremberg's *Historia naturae, maximae peregrine* (1635) and the Academy of the Lynx's *Rerum medicarum Novae Hispaniae thesaurus* (1651). But López Piñero and his Valencian colleagues have demonstrated that many copies of different sections of Hernández's manuscript had long been circulating in Mexico, Spain, Holland, and Britain. Pieces of Hernández's labor surfaced in works by Gregorio López (ca. 1583; published 1678), Juan Barrios (1607), Francisco Ximénez (1615), Johannes de Laet (1625, 1630, 1633), Georg Margraf (1648), Robert Lovell (1659), Henry Stubbe (1662), Hans Sloane (1707–25), James Newton (1752), and James Petiver (1715). We also know now that some of the original illustrations of Hernández's natural history survived in such manuscripts as the Codex Pomar.[54]

Two handsome volumes on Hernández published by Stanford University Press, *Searching for the Secrets of Nature* (2000; ed. Varey et al.) and *The Mexican Treasury* (2001; ed. Varey), go a long way toward making this new Spanish scholarship available to English-speaking audiences. *The Mexican Treasury* offers readers a thoughtful selection of some of Hernández's writings. It includes pieces drawn from his extant manuscripts, as well as selections from several of the authors who copied his writings in the course of the seventeenth and eighteenth centuries. Although the selection privileges Hernández's natural history, it also includes a translation of his correspondence with Philip II and Arias Montano, his will, his "Christian Doctrine" (a pedagogical poem he wrote while in Mexico summarizing the tenets of Catholic theology), and excerpts from his treatise on Mexican antiquities. Although these selections clearly encourage contextual readings of his work, they do not go far enough. We know, for example, that Hernández translated and glossed Aristotle, Pliny the Elder, and Pseudo-Dionysius. He also wrote essays on Stoic philosophy. Samples of these writings should have been included as well.

The accompanying volume, *Searching for the Secrets of Nature*, seeks to put Hernández in his appropriate historical context. There are a few essays in this collection that are particularly insightful. Peter O'Malley Pierson presents Philip II not as an obscurantist monarch but as a patron of natural philosophers whose generosity was always limited by a bankrupt treasury. Rafael Chabrán locates Hernández in the philological and experimental traditions of Spanish humanism that thrived at the universities of Alcalá de Henares and Valencia. Hernández found in Nahuatl etymologies and taxonomies an alternative to the botanical classifications of Dioscorides. He also sought to confirm the medical virtues of plants through clinical trials. Guenter B. Risse offers an enlightening study of sixteenth-century Mexican hospitals, where local shamans introduced Hernández to Nahuatl botanical knowledge, and where Hernández did his clinical research. Essays by López Piñero and Pardo Tomás and by Rafael Chabrán and Simon Varey painstakingly reconstruct the history of the dissemination of Hernández's manuscripts in Mexico, Spain, Britain, and the Netherlands. Jaime Vilchis highlights the importance of understanding Hernández's Neoplatonic and Stoic writings to comprehend why he went to Mexico in the first place. Hernández still remains a poorly understood figure. Yet these two books have contributed to lifting the fog enveloping his life and work. The contribution in *Searching for the Secrets of Nature* that truly breaks new ground is that by María José López Terrada, who studies the incorporation of new plants of the Indies into the repertoire of Golden Age painting. López Terrada demonstrates the awareness of Spanish painters of the botanical novelties introduced to the Iberian Peninsula by the writings of Nicolás Monardes and Hernández; more important, his article points to the impact that New World flowers and plants had on the early modern garden.

The history of Spanish gardens cannot be overlooked in a review of the literature on early modern Spanish science. The Renaissance garden expressed the values and aspirations of natural philosophers. Like the medieval gardens of courts and cloisters, early modern gardens sought to re-create paradise.

Fountains in Catholic cloisters stood for the blood of Christ, the giver of life; the enclosing walls symbolized the virginity of Mary and the spiritual purity of the Church; flowers represented virtues, and weeds, vices.[55] Humanism added pagan and utilitarian dimensions to the mix. In the fifteenth and sixteenth centuries, the medieval garden, a locus of contemplation and amorous courtship, was transformed into a space that promoted mastery and dominion over territory.[56] It also became a place to store and reproduce encyclopedic knowledge (particularly in the botanical garden);[57] to reproduce the geometrical Neoplatonic structure of the cosmos;[58] and to build patriotic replicas of local landscapes.[59] Visitors, like epic heroes, stepped into mazes and labyrinths on voyages of philosophical discovery.[60] Marble sculptures, topiary work, grottos, and loggias helped these voyagers find their cues along the way. The study of Renaissance and mannerist Italian, French, English, and Netherlandish gardens as microcosms of early modern knowledge is a well-established field. Yet, typically, scholars have paid little attention to developments in Spain.

Carmen Añón Feliú has long been seeking to correct this oversight.[61] *Jardín y naturaleza en el reinado de Felipe II* (Garden and Nature in the Reign of Philip II), a book she edited in 1998 along with José Luis Sancho, demonstrates, in case there was any doubt, that sixteenth-century Spain was part of the much larger culture of the European Renaissance, and that many gardens, old (Arabic) and new, flourished under the leadership of Philip II: in the Reales Alcázares of Seville, the Alhambra, the Royal Garden of Valencia, the Casa de Campo, El Pardo, Aranjuez, Vaciamadrid, Valsaín, Escorial, La Fresnada, El Quexigal, and the *dehesas* of Campillo and Monasterio. Like their English, Italian, and French peers, Spanish scholars imagined and created philosophical gardens and embarked on the construction of allegorical and emblematic landscapes. More important from the perspective of this review, natural philosophers used the garden to encapsulate the new encyclopedic knowledge arriving from the Indies. Gardens were part of the larger set of institutions and practices put in place to master the botanical resources of the Indies.

This newfound interest in the role of botany in the history of early modern Spanish science and culture is part of a larger discursive pattern in Spanish scholarship. In *La conquista de la naturaleza Americana* (1993), Raquel Alvarez Peláez explores many of the same authors and developments that captured the attention of López Piñero and his Valencian colleagues throughout the 1990s. Like them, Alvarez Peláez seeks to understand the Spanish contributions to early modern European botanical and medical knowledge, but she brings in new sources to the discussion, namely, the hundreds of documents from the Spanish American colonies sent to answer the "geographical" questionnaires ordered by Philip II in 1577 and 1584. These so-called *relaciones geográficas* have long attracted the attention of historians and ethnohistorians, for they contain useful information on the social, cultural, and economic history of dozens of regions in the Americas. Cities, towns, and pueblos across the continent scrambled to reply to the queries by submitting maps, narrative histories, and detailed descriptions of local natural resources. Alvarez Peláez studies some fourteen volumes' worth of replies from Mexico alone for evidence of contemporary understanding and mastery of natural history. The conclusions she reaches are not particularly surprising: Spanish authorities all over the continent duly replied to the metropolitan queries by imposing European categories on local resources, while thoroughly relying on local informants and knowledge.

Also not surprising is what moved Alvarez Peláez to write this lengthy study. Throughout, she seeks to prove that the charges leveled against Spain by Antonello Gerbi in his celebrated study *La natura delle Indie nove: da Christoforo Colombo a Gonzalo Fernández de Oviedo* (1975) are wrong, for Gerbi had argued that, compared to the Italians in the sixteenth century, Spanish naturalists in the Americas clearly lacked descriptive and analytical powers. According to Gerbi, Fernández de Oviedo was the exception, largely because he had been trained in Italy. Alvarez Peláez rejects this characterization.[62] Like her peers in the seventeenth and eighteenth centuries, Alvarez Peláez takes up her pen to battle foreign innuendos. Claims to objective detachment notwithstanding, patriotism remains a powerful motivating factor in modern Spanish scholarship.

Patriotism can engender creative and stimulating work. Alvarez Peláez's book, however, is not particularly illuminating. She uses the rich documentary cache of the *relaciones geográficas* to reconstruct some Mesoamerican indigenous medical and botanical categories, but without much anthropological imagination. For a more inspiring example of what can be done, the reader should peruse Barbara Mundy's *The Mapping of New Spain* (1996). Using the same sources, Mundy offers a dazzling study of the rise of hybrid conceptions of space and mapmaking in sixteenth-century Mexico.

Like Alvarez Peláez's book, José Sala Catalá's posthumous treatise *Ciencia y técnica en la metropolización de America* (1994) is a study of the history of Spanish science in colonial settings. Like Alvarez Peláez, Sala Catalá also has a patriotic agenda, namely, to demonstrate the scale, complexity, and sophistication of the Spanish sixteenth- and seventeenth-century technologies that went into the building of the new cities of Mexico and Lima. The draining of the central valley of Mexico alone, Sala Catalá concludes, constituted "the most important piece of civil engineering of the Renaissance." Mexico, the author forcefully maintains, became "the most extraordinary laboratory of hydraulic experimentation [*experiencias*] in the world."[63] Readers who discount these assertions as mere patriotic nonsense do so at their own peril. Persuasively and ingeniously, Sala Catalá reconstructs the multitude of complex technologies involved in the Spanish urban settlement of America.[64] His long and beautifully illustrated account of the feats of hydraulic engineering required to drain the central valley of Mexico by cutting sluices through the massive mountains surrounding the valley is particularly illuminating. It shows among many other things the multinational character of the empire, for Flemish technicians first directed these works.

The chronic Spanish reliance on foreign technicians is a theme that runs through the British scholar David C. Goodman's *Power and Penury: Government, Technology and Science in Philip II's Spain* (1988).[65] This book is as comprehensive and synthetic as López Piñero's *Ciencia y técnica* and García Tapia's *Técnica y poder*. In fact, Goodman covers remarkably similar ground. *Power and Penury* describes in encyclopedic fashion developments in science and technology ranging from alchemy to cosmography, shipbuilding to artillery, metallurgy to navigation, and surgery and medicine to botany. Goodman discovers for himself something Spaniards had been saying for centuries: that Spain was intellectually part of Europe. Philip II was not a superstitious, ignorant tyrant but a monarch remarkably open to innovation in new experimental fields. His palace at El Escorial contained a large alchemical laboratory to produce medicinal waters by distillation, as well as the much-hoped-for transformation of base metals into gold and silver. Like other monarchs of his age, Philip believed in astrology and the occult, but he remained far more skeptical than many of his European colleagues. Although he sought to control and regulate the training of ships' pilots, gunners, physicians, and surgeons, and although the shadow of the state lurked behind most of the great technological and scientific developments of the age, there was great room in Spain for private initiative and entrepreneurship. Spaniards were no more adverse to science and technology than other European peoples. What ultimately kept many away from technical professions were low wages, not any innate aristocratic aversion to menial occupations. For all his goodwill, Philip II was always plagued by chronic fiscal deficits, unable to sustain long-term initiatives and support foreign technicians.

That all these points needed to be made by Goodman in 1988 is a testament to the power of the traditional narratives of the Scientific Revolution that located the roots of modernity in astronomy and physics, not natural history. The marginalization of the epistemic traditions first introduced by the early modern Iberian empires to the genealogies of modernity is also a testament to the power of the printing press. The early modern Spanish Crown held knowledge to be a state secret (*arcana imperii*), and it therefore circulated through manuscripts. It has taken historians many years, if not centuries, to unearth the evidence needed to upset the well-established narratives of the Black Legend of Spanish backwardness. These prejudices are still with us, blinding historians every day. New accounts of the Scientific Revolution by Steven Shapin (1996) and Peter Dear (2001) manage to exclude the Iberian empires as completely as Marie Boas and A. Rupert Hall once did. It is extraordinary that decades after the publication of José Antonio Maravall's *Antiguos y modernos* (Ancient and Moderns) and López Piñero's *Ciencia y técnica*, North Atlantic historians of early modern science and technology still can so easily elect not to study this literature. Fortunately, as demographic patterns in the United States change, the self-satisfying North Atlantic narratives of the origins of modernity are in for a rude awakening. It is just a matter of time before books in English on the Scientific Revolution begin adorning their dust jackets with the frontispiece of García de Céspedes's *Regimiento de navegación*, rather than that of Bacon's *Instauratio magna*.

CHAPTER 3

From Baroque to Modern Colonial Science

Formal systems of science created by the Aztecs, Maya, and Inca seem to have evaporated into thin air in the wake of the Spanish conquest.[1] The collapse of large indigenous polities and the disappearance of courts capable of sustaining elite knowledge appear to be the causes. Nancy Farriss has argued that the Maya in Yucatan lost the institutions that had kindled their taste for large cosmic riddles. Although the Maya elites did not disappear—and actually became important brokers in the operation of colonial labor systems—they lost interest in the theological and cosmological questions that had driven the astronomical and calendrical investigations of the classic and post-classic Maya civilizations. As the Maya elites were left in charge of ever more simplified polities, their interest became narrowly parochial. Under Spanish colonial rule, the former complex social structures of the Inca, Maya, and Aztecs gave way to simplified communities lacking all intermediate social tiers: gone were the indigenous pan-regional polities of the past whose courts had maintained large retinues of priests, scribes, and scholars—producers of elite precolonial non-Western knowledge. The new simplified native elite class embraced Catholic images, shrines, temples, and rituals, and those few religious leaders who kept native religions (and thus non-Western scientific traditions) alive went underground, losing the source of much of their prestige, which lay in maintaining communal cohesion through sumptuous *public* worship. By the eighteenth century, indigenous systems of knowledge had transmuted into hybrid forms of folk Catholicism and had become marginal in Latin American societies.[2]

This chapter does not deal with the hybrid forms of popular knowledge that developed on the fringes of colonial Latin American societies, such as those produced by the millions of African slaves who arrived in Spanish and Portuguese America over the course of four centuries. The history of science of colonial Latin America, by and large, does not belong in the "non-Western world." The scientific practices and ideas that became dominant were those brought by Europeans as they strove to create stable, viable colonial societies. Portugal failed to introduce learned institutions in Brazil until the early nineteenth century, however, when, in the wake of Napoleon's invasion, the Portuguese Crown fled Lisbon to settle in Rio de Janeiro. This chapter focuses primarily on the viceroyalty of New Spain. In the eighteenth century, Mexico produced most of the wealth Spain derived from its colonies. Mexican elites were wealthy and cosmopolitan; in 1734, for example, the cathedral cloister in Mexico City inaugurated one of earliest standing orchestras of the "Western world."[3] Eighteenth-century Mexico is ideally suited for a study of the connections of science to baroque culture (which was characterized by an emblematic view of nature), to colonialism, and to nationalism.

Early Institutions

In the early eighteenth century, the Spanish Crown appeared to be saddled with highly autonomous colonial societies. Although for some two centuries the New World had supplied Europe with silver, sugar, and dyes, Spanish America was hardly just a colonial outpost solely serving the needs of the metropole. The loose political structure of the Spanish empire, Spain's sluggish markets, and the somewhat limited demand for colonial products had combined to turn the colonies into semi-independent entities ruled by viceroys and *audiencias* (high courts) that were always observant of the needs of local elites.

Like any other early modern European society, Spanish America was founded on corporate privileges and social estates. Yet unlike other European societies, Spanish America had social estates that overlapped with racial and other cultural hierarchies: African blacks were slaves; "Indians" were treated as peasant commoners and regarded for legal purposes as childlike members of a separate "republic"; and Spaniards and their descendants enjoyed special privileges and thought of themselves as patricians. *Castas* (mixed bloods) lived in the interstices of the original three-tier system, and as they blurred the carefully policed colonial racial boundaries, they gave way to complex social taxonomies and intermediate groups of commoners.

By the early eighteenth century, Spanish America had developed a set of institutions that fostered scientific activities: universities and colleges, monasteries and convents, private libraries, pharmacies, and viceregal and ecclesiastical courts. Spanish America boasted some twenty universities and dozens of religious colleges, chartered on the model of the medieval University of Salamanca and controlled by religious orders (mostly Dominicans and

Jesuits), which trained theologians, lawyers, and a few physicians in neo-scholastic curricula.[4] The Jesuits developed powerful educational institutions of their own that catered to the needs of local elites. Their support for philosophical eclecticism allowed the followers of St. Ignatius Loyola to introduce their charges to some innovative European thought, including experimental philosophy. The Jesuits subordinated science to their apostolic mission and created vertically integrated and technically efficient economic institutions (e.g., haciendas and plantations) to support their colleges and missions. The operation of their pharmacies is a case in point. Luis Martín has argued that in the seventeenth and eighteenth centuries, the Jesuits ran a network of pharmacies in various colonial cities from their college of San Pablo in Lima. Italian and, later, German brethren were charged with making the pharmacies profitable, and they set up labs and collected and exchanged plants in hopes of identifying new remedies for trade. According to Martín, the Jesuit pharmacy in Lima held a monopoly on quinine and bezoar stones in European markets throughout the seventeenth and early eighteenth centuries, and the profits helped the order to maintain missions and colleges.[5] The Jesuits also subordinated cartography and natural history to the strategic needs of the order, which sought to expand post-Tridentine Catholicism to the frontiers of the colonial Spanish empire. From 1628 to 1767, the year in which they were expelled from all Spanish possessions, the Jesuits were the officially appointed cosmographers of the Indies.[6] The order promoted coordinated astronomical observations and the writing of natural histories.[7]

Viceregal courts and convents offered alternative patronage systems to that of the universities, allowing some innovative philosophers to emerge. Baroque polymaths kept cabinets of curiosities and alchemical laboratories, and they were summoned by patrons to make astronomical observations and draw maps, to cast horoscopes, to design machines for courtly and public entertainment, to tend to the sick, and to help design sacred and secular public buildings. Carlos Sigüenza y Góngora best typifies the Spanish American baroque polymath. Holder of the chair of mathematics at the University of Mexico, Sigüenza taught medical astrology, while also working as a censor for the Inquisition and as chaplain for a local hospital. He drew maps, helped coordinate the work to drain the lake on which Mexico City was built, and led a surveying expedition to the borderlands (to the bay of Pensacola in Florida). Sigüenza was an accomplished scholar who kept a cabinet of curiosities and a telescope and a microscope and who did not fear to engage in heated debates with European astronomers. In 1690, Sigüenza published his *Libra astronómica* to take on the German Jesuit Eusebio Francisco Kino,

former professor at the University of Ingolstadt and a leading missionary in
California, for having espoused antiquated theories on the origin of comets.
In his argument that comets were not "earth exhalations" (ground vapors that
congealed in the sublunar sphere), harbingers of disease, and omens of evil,
as German and other astrologers had long argued, Sigüenza showed famil-
iarity with the writings of Kepler, Galileo, and Descartes. Finally, Sigüenza
spent a great part of his scholarly life seeking to clarify biblical chronologies
through a detailed study of Mesoamerican calendars and codices.[8]

More often than not, baroque polymaths were summoned to participate
in the rituals of power that facilitated the smooth reproduction of social
structures.[9] *Cabildos* (city councils), monastic establishments, viceroys, and
prelates supported savants to design emblems for triumphal arches and
funeral pyres, to build machines to impress the public in religious proces-
sions, to write and deliver commemorative sermons, and to uncover hidden "signatures" in nature and religious images. It is no wonder therefore
that Neoplatonic and hermetic currents enjoyed wide currency among the
learned. The Neoplatonic writings and theories of the seventeenth-century
authors Juan Caramuel, Athanasius Kircher, and Caspar Schott had a particularly strong and lasting influence on the imaginations of most baroque
Spanish American scholars.[10] In fact, Kircher himself exchanged letters with
many a Mexican scholar, to whom he sent academic advice, books, religious
images, and mechanical toys in exchange for curiosities, patronage, and dozens of pieces of prime Mexican chocolate (which he seems to have loved).[11]
Kircher was so grateful that he dedicated one of his treatises on magnetism
to one of his Mexican correspondents, Alejandro Favián.[12] Sigüenza typifies
the Spanish American baroque polymath collaborating in the theatricaliza-
tion of power. He wrote commemorative pamphlets on civic and religious
events and designed emblems for triumphal arches. Drawing on cosmic
metaphors, he readily represented secular and religious authorities as "suns"
and "planets" of a hierarchical sociopolitical order. He corresponded with
Caramuel, and his will indicates that the tooth of a "giant," a collection of
Mesoamerican codices, and the works of Kircher were among his dearest
possessions. Sigüenza donated his body to be dissected.[13]

Patriotic, Neoplatonic, and Emblematic Dimensions

Colonial baroque scholarly traditions were strongly colored by patriotism. As
David Brading has shown, by the seventeenth century, "Creole patriotism"
had penetrated most learned circles in the colonies. Creoles, those born

in the colonies of Spanish descent, saw themselves as being discriminated against by first-generation Spanish migrants. Creoles took the newcomers to be lowly commoners whose activity in commerce and mining had given them access to landed wealth and false claims to patrician origins. Creoles were in turn perceived by peninsular Spaniards as idle dilettantes who had turned into degenerate Amerindians, either because of astral influences or cultural proximity to the natives. Although the Crown privileged peninsulars in colonial appointments, Creoles took advantage of Spain's chronic fiscal crisis to buy offices, and they were thus kept out of only the highest colonial posts. Creoles secured positions in the Church and gained power over local ecclesiastical institutions and religious orders. Creoles also thought of themselves as a local landed nobility. But since colonial landed elites were never allowed to become feudal lords or grandees, thus increasing land turnover and continuous division of land among heirs and lowering the incentive to invest in and improve landholdings, Creoles tended to be downwardly mobile and resented the success of Spanish newcomers, who paradoxically aspired to gentility by marrying into Creole families. As learned clergymen, Creoles wrote patriotic treatises praising the glories of the land, themselves, and their ecclesiastical establishments (monastic institutions, churches, nunneries, and universities).[14] Natural philosophy in the colonies was also patriotic.

Steeped in Neoplatonic and hermetic doctrines, the Creole clergy were constantly searching in nature for underlying hidden signatures with patriotic significance. For them, the human body, the Earth, and the cosmos were all baroque "theaters" (in that objects were reduced to a language of images) interlocked by micro- and macrocosmic analogies.[15] All objects held polysemic meanings, and the exegetical skills of the clergy helped discover their underlying import, revealing a cosmos suffused with providential designs that favored the colonies. For example, Creole scholars concluded that astral phenomena that would have caused natural disasters in Europe were benign in the Indies. In 1638, the Augustinian friar Antonio de la Calancha (1584–1654) argued that eclipses in Asia, Africa, and Europe were ominous signs; should they happen under Aries, Leo, or Sagittarius, they would set off horrible aerial visions, harmful comets, and devastating fires; under Gemini, Libra, and Aquarius, they would trigger famines and epidemics. According to Calancha, Peru had so many new stars, however, that it was the domain of entirely different and auspicious signs of the zodiac. The five-star constellation of Cruzero (the Southern Cross), for example, was at the southern pole, and its crosslike form kept away demons responsible for stirring up the waters. It was for this reason, he argued, that the South Seas were

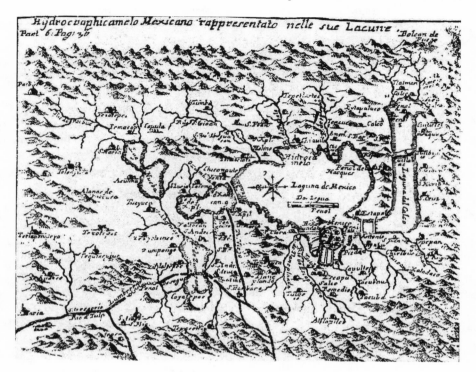

FIG. 3.1. "Hydrocoophicamelo [*sic*] Mexicano rappresentato nelle sue lacune" ("Hydrographicamel" of Mexico Represented in Its Lakes), from Giovanni Francesco Gemelli Careri, *Giro del mundo* (Naples, 1699–1700). The map was first drawn in 1618 by Adrian Boot, a Flemish engineer hired to oversee the works leading to the drainage of the central valley of Mexico. The figural reading of the map, however, appears to be by the mathematician and cartographer Cristóbal de Guadalajara. Along with the map, Gemelli published cabalistic readings of the names of the ten Aztec monarchs showing that their numerological value added up to 666, the number of the Beast of the Apocalypse.

calm and were called the "Pacific" Ocean. Calancha thought, in fact, that Peru was blessed; God had chosen to protect it by giving it not only cross-like constellations but also crosslike fossils, stones, and plants.[16] In Mexico, things were slightly more complex, for demons appeared to be so entrenched there that they had even shaped the landscape. In the 1690s, Cristóbal de Guadalajara, a mathematician and cartographer from Puebla, realized after studying a copy of an early seventeenth-century map of the rivers and lakes of the central valley of Mexico that the hydrographic contours of the valley represented the head, body, tail, horns, wings, and legs of a Satanic beast.

To counteract the Aztec kingdom of darkness, God had fortunately sent Spanish conquerors and the Virgin Mary to set the Amerindians free. The Virgin—who had appeared to a Nahua commoner, Juan Diego, in Tepeyac in 1531, leaving her image stamped on his cape—held the key to understanding Mexico's destiny. The image, a variation of a rather common European representation of the Immaculate Conception, had a Virgin standing on the moon, surrounded by sun rays while eclipsing the sun, wearing a heavily starred blue shawl, and held up by an angel (fig. 3.2). In 1648, Miguel Sánchez argued that the woman of the Apocalypse, pregnant with a potential messiah, clad in stars, threatened by a multiheaded dragon (the devil), who wants her to miscarry, and protected by God, who sends an army of angels headed by the archangel Michael to destroy the dragon (Rev. 12:1–9), is a prefiguration of Our Lady of Guadalupe, who had routed the kingdom of darkness of the Aztecs. Sánchez offered interpretations of every detail of the image: the moon underneath the Virgin represented her power over the waters; the Virgin eclipsing the sun stood for a New World whose torrid zone was temperate and inhabitable; the twelve sun rays surrounding her head signified Cortés and the conquistadors who had defeated the dragon; and the stars on the Virgin's shawl were the forty-six good angels who had fought Satan's army (Sánchez used cabalistic methods to calculate the number of good angels).[17] Sánchez thought that the image was the most important icon in Christendom, and his interpretation inaugurated a literature of exegesis in which contemporary Mexicans appeared as God's new elect.[18] Throughout the seventeenth and eighteenth centuries, baroque Creole scholars debated the meanings of the image and reached conclusions such as that it had dominion over the sphere of water, because the Virgin was standing over the moon, which, in turn, was related to tides, floods, and droughts.[19] Every time Mexico City was flooded (which happened often), thousands of anguished citizens took to the streets to parade her image, and to their relief, the waters always subsided.[20] If properly understood, religious images could be deployed at cardinal points of cities and towns to act as symbolic "fortifications" (baluartes), to fend off evil sublunar "intelligences" capable of causing natural calamities.[21]

Latin American historians of science have not given proper attention to the emblematic and Neoplatonic dimensions of colonial science and have misread key colonial baroque texts. Elías Trabulse, a prolific Mexican historian of science, has shown the Mercederian Diego Rodríguez to be a semi-

FIG. 3.2. Frontispiece to Cayetano de Cabrera y Quintero's
Escudo de armas de México (Mexico City, 1746). The
image of Our Lady of Guadalupe is supported by putti and
acts as a shield protecting Mexico from negative heavenly
influences, reportedly the underlying cause of the epidemics
of *matlazahuatl* that hit Mexico between 1736 and 1738.
The image of the Virgin acts as a "fortification" to keep evil
intelligences away.

nal Creole natural philosopher, the first holder of the chair of mathematics at the University of Mexico and the author of highly sophisticated mathematical treatises. The merit of Trabulse's work on Rodríguez is that it puts to rest the rather popular construct that Catholic colonial Spanish America was intellectually barren, a land choked by the Inquisition. Trabulse presents Rodríguez as a "modern," whose treatise on the comet of 1652 sought to dispel "superstitious" beliefs that held that comets were harbingers of evil.[22] Trabulse has insisted correctly that Rodríguez showed an acquaintance with the works of Kepler, Galileo, and Descartes. But Rodríguez was a Creole patriot engaged in Neoplatonic and emblematic readings of nature. In fact, Rodríguez's treatise shows that his rejection of comets as ominous signs stemmed from his repudiation of a common idea held by Europeans that the constellations of the New World were different from those of the Old World and caused biological degeneration; his patriotic belief that Mexico's skies were protected by the Immaculate Conception; and his conviction that "there is no sign [over the skies of the viceroyalty of New Spain] that, although shocking and surprising for the ignorant, does not serve the Queen of Heaven and [help] explain her glories."[23] The basic assumption that moved Rodríguez to deny that the comet of 1652 was a harbinger of disease and death was that its path through the zodiac revealed its symbolic associations with the Immaculate Conception. The comet had moved through the constellations of Noah's Dove (Columba) and Medusa (?). The constellation of the dove stood for purity, not unlike that of Mary, whose immaculate conception had spared her a postlapsarian human nature; that of Medusa, on the other hand, represented the dragon that had sought to kill the pregnant Virgin. Rodríguez assumed therefore that the comet and the Virgin Mary were symbolically linked. Rodríguez also presupposed that since the image of the Immaculate Conception was a Virgin eclipsing the sun, eclipses as well as any other heavenly phenomena in Mexico, a land under the protection of Our Lady of Guadalupe, could be harbingers only of joyous news. According to Rodríguez, the comet of 1652 would make Mexican leaders wiser, because it had passed across Mars, a planet that stood for wisdom. To be sure, his argument for the benignity of the comet was not sustained solely on the learned exegesis of nature and religious images. Based on the allegorical interpretation of classical mythology as physical events and on close knowledge of contemporary astronomy, Rodríguez sought to prove that comets were supralunar phenomena and therefore had no negative physical effects on the sublunar world. According to Rodríguez, the sub- and supralunar worlds were qualitatively different. Emblematic readings of nature lasted well into the eighteenth century. In 1742, for example, the Augustinian friar Manuel Ignacio Farías sought to explain the cause of a bout of *matlazahuatl* (plague) that had killed thousands of indigenous people of Michoacán. Like Rodríguez, Farías thought that the image of the Immaculate Conception and heavenly phenomena in the Americas were symbolically interwoven and that the skies of Mexico were under the special patronage of the Virgin. Empirical evidence proved, however, he believed, that the epidemics had occurred in the wake of a solar eclipse. If it was true that the image of Our Lady of Guadalupe protected Mexico, the recent eclipse that had been observed in Michoacán should never have triggered such natural disaster, for the image showed the Virgin eclipsing the sun. Calancha and Rodríguez had already argued that eclipses in the New World could be harbingers only of joyous news. Farías thus regarded the recent event as an anomaly, which had been prefigured in the Bible. Interpreting words of King David as predicting that those standing to the right of the Temple would never be protected, Farías maintained that the natives of northern Michoacán lived to the "right" of the shrine of Guadalupe in Valladolid (capital of Michoacán), for the image of Our Lady of Guadalupe in the church was facing west to replicate on a larger scale the message of the painting (that the image should "eclipse" the sun at sunrise to act as a protective shield for the city). Given the logic of Farías's cosmology, those who "stood" to the "right" of the image were the Amerindians of the north.[24]

These kinds of emblematic, sacred, and patriotic readings of natural phenomena were often used in medical treatises. Eighteenth-century Creole doctors concluded, for example, that pulque, a liquor from the maguey plant, was a panacea, not simply because clinical evidence indicated this, but also because maguey was the plant from which the fabric on which the miraculous image of Our Lady of Guadalupe was stamped had been made; the Virgin had subtly let medical doctors know about the divine virtues of maguey.[25] Many of these physicians thought that the virtue of plants could also be identified by studying Nahua etymologies. In 1746, Cayetano Francisco de Torres argued that pulque was a universal remedy (*polychresto*) because it had symbolic-material association with the image of Our Lady of Guadalupe and because the Aztec name of the plant, *teometl*, "divine plant," clearly said so.[26]

The idea that Nahuatl and Quechua, the tongues of the Aztecs and the Inca, were "Adamic" languages exerted an important attraction on the imaginations of Creole scholars in Spanish America throughout the entire colonial period. In their efforts to claim the status of independent "kingdoms," endowed with prestigious genealogies and loosely affiliated with a universal Spanish monarchy, for their homelands, some Creoles converted the indigenous past into their own classical antiquity. Many Creole scholars were therefore willing to compare the languages of the Aztecs and Inca favorably to Greek and Latin. And some naturalists even claimed that Nahua and Quechua taxonomies pointed to the essences of plants.

The naturalist Francisco Hernández, commissioned by Philip II to identify the properties of plants in the New World, was perhaps the first to suggest, circa 1574, that Nahuatl, the language of the Aztecs, was an Adamic language. He insisted that, like the ancient Hebrews, Mesoamerican peoples had named things after their essences. He expressed surprise that nations he thought were barbarous could have developed such sophisticated languages.[27] In 1637, on the advice of the holder of the chair of Quechua (the Inca language), Alonso Huerta, the faculty of the University of Lima turned down a proposal to establish a new chair of medicine devoted to botanical studies, on the grounds that physicians should rather study Quechua. Peruvian Amerindians, the faculty argued, had already discovered the properties of plants and had named them after their virtues.[28] In the eighteenth century, some Peruvian Creole naturalists insisted that new educational institutions devoted to the study of nature in the colonies educate students in Quechua,[29] and we know that important late eighteenth-century naturalists in Lima, such as Juan Tafalla and Francisco González Laguna, who were deeply involved in the royal botanical expedition to Peru of 1777–1808, also sought to promote the publication of Quechua grammars.[30] In Colombia and Mexico, prestigious Creole naturalists such as Francisco José de Caldas and José Antonio de Alzate openly praised the value of Quechua and Nahua taxonomies.[31] Finally, the Spaniard Martín de Sessé, head of the botanical expedition to Mexico in 1787–1803, made it his priority to learn Nahuatl, on the assumption that the language of the Aztecs was "an elegant language [in which the] names of plants signified their virtues."[32]

In Service to Crown and Commerce

By the mid eighteenth century, these hierarchical yet fiercely patriotic baroque societies had been changed by forces outside their control. The Spanish Crown embarked on a project of economic and cultural renewal and set out to reform outdated policies, revitalize Spanish commerce, make Spain a European power again, and prove that it was not intellectually barren, as foreign sarcasm held. To work, the program of reform needed to turn Spain's

overseas territories into colonial dependencies, politically and economically subservient to the needs of the metropole. It also required that science
be enlisted in the service of the state and the new economy.[33] Science
would provide the Crown with new secular courses of political legitimacy
to undermine the power of the Church.[34] Natural history, experimental
philosophy, astronomy, and cartography would help the Crown to exploit
botanical and mineral resources and to regain control over loosely controlled frontiers and borderlands.[35] Under the aegis of Charles III, Spain's
revitalized navy conveyed numerous expeditions staffed by *peninsulares* and
other Europeans to America to draw accurate maps of territories in danger
of being lost to England, Portugal, France, and Russia, as well as to identify
botanical resources in order to establish new commercial monopolies.[36]
Much energy was wasted by numerous bureaucrats and naturalists bent on
finding cloves, cinnamon, and tea in tropical America to challenge the monopolies of Dutch and British merchants.[37] Spanish botanical expeditions in
America also sought to tap the continent's vast and unknown pharmaceutical
resources to find "a panacea for the diseases of the century."[38] Finally, the
Crown also sent out parties of Spanish and German chemists, geologists,
and mining experts charged with improving the production of silver and
mercury.[39]

Spain obtained access to the new western European sciences by sending
students abroad, by hiring foreign technicians and savants, and by creating
new institutions of learning both at home and in the colonies.[40] The Crown
sought to reform universities to gain control of the Church. Many Jesuit
educational institutions were dismantled when the order refused to embrace
the new regalist and Jansenist principles of relations between Crown and
Church, and the Jesuits were expelled from all Spanish territories in 1767.
The Crown seized the momentum of the Jesuit expulsion to undermine
the autonomy of all universities and to introduce reforms. Yet the universities did not become sites where the new science set roots, for their faculties
understood correctly that the new learning was associated with attempts to
undermine their corporate privileges and to terminate the clerical monopoly
on learned institutions.[41]

Instead, as Antonio Lafuente and José Luis Peset have argued, Spanish science of the eighteenth century was modernized by the military.[42] The army
and the navy created a set of alternative establishments to the universities
where state-of-the-art sciences were taught, including mechanical philosophy, Newtonian physics and the calculus, and experimental philosophy. The
military set up hospitals in Spain and in the colonies for learned surgeons,

mathematical academies in Spain for gunners and engineers, and an observatory in Cadiz for cartographers, as well as helping support a network of botanical gardens to acclimatize tropical plants that might prove useful for new commercial ventures. But the Crown also sponsored new institutions that were independent from the military. Academies of art were created in Spain and Mexico to train masons, architects, textile designers, and botanical illustrators in neoclassical taste.[43] The Royal Academy of History (1738) in Madrid and the Archive of the Indies (1784) in Seville were founded to bring critical methods to bear not only on the study of historical documents but also on geography and natural history (one of the most important responsibilities of the Academy of History was the writing of critical natural histories). And medical academies, the Royal Botanical Garden (fl. 1774–88), and a revived *protomedicato* worked in Spain and in the colonies to standardize the training of healers, to undermine guilds of apothecaries and surgeons, to check quacks and midwives (and shamans in the case of the colonies), and to improve the lowly status of physicians. Like other European monarchies, the Spanish Crown linked mercantilism to medical reform in the hope of augmenting the population, and it avidly endorsed large-scale smallpox immunization at home and in the colonies. The Crown also supported the development of a public sphere throughout the empire and promoted salons and patriotic societies to help disseminate the new utilitarian learning.[44]

But the mercantilist project proved unsuccessful in the long run. The French Revolution sent chills down the spine of the Spanish monarchy, whose active involvement in the creation of a public sphere in Spain was suddenly checked, freezing most initiatives for cultural renewal. Continuous war against England threw a monkey wrench into the efforts of the Crown to reorganize colonial trade. The frustrated policies of economic and cultural renewal did not leave colonial science unscathed. Francisco Puerto Sarmiento has described how the institutions created by the Crown to exploit colonial botanical resources slowly deteriorated, leaving Spain with little to show after some thirty years of unparalleled official patronage of naturalists. Botanical gardens set up to acclimatize tropical plants were shut down or never used; the efforts to standardize the production of quinine to take advantage of Spain's secular monopoly on the febrifuge failed; production of cinnamon and cloves never succeeded; and, most ironically, the patriotic evidence that could at last have proven to the rest of Europe that Spain was indeed "modern"—taxonomies, travel accounts, maps, and thousands of botanical and anthropological illustrations accumulated by dozens of scientific expeditions—remained, by and large, unpublished.[45] The institutions and expeditions created to increase the productivity of silver and mercury mines in Mexico and Peru failed to introduce significant technological changes; and although production of silver increased greatly, this was largely a result of the extensive use and overexploitation of manual labor.[46] Moreover, as Juan José Saldaña has demonstrated, the Colegio de Minería (College of Mining), established in Mexico by the Crown in 1792 to churn out bureaucrats and miners trained in subjects such as geometry, geology, and chemistry frequently was not funded; it remained a vibrant institution only because the local elites chose to keep it alive.[47] Medical reform in the colonies did not improve public health significantly, nor did reform elevate the social status of physicians.[48]

But despite all these failures, the cultural milieu of the colonies was forever changed in the wake of the Bourbon efforts at imperial renewal. A public sphere came into being, for one thing. Newspapers, periodicals, salons, cafes, and patriotic societies appeared everywhere, divulging new European thought and the gospel of utilitarian knowledge. Eventually, this dynamic public sphere was to demand new forms of democratic political participation from the Crown.[49] "Newtonianism" and mechanical philosophy permeated most public discourse. Even opponents of the new science (who correctly perceived that this new learning was peripheral to clerical institutions and was contributing to the secularization of society and the demise of scholastic theology) studied it.[50] The rhetoric of experimentation and the fad of collecting cabinets of experimental apparatus overtook the learned.[51] Desiderio de Osasunasco set up a most elaborate experimental plan in the 1780s to discover the cause of his own intolerance to chocolate, subjecting himself to systematic and painful self-experimentation. The doctor Juan Manuel Venegas thought little of the natives, whom he considered ignorant savages, yet he wrote a medical treatise on indigenous herbal lore on the assumption that these savages knew things discovered through a process of trial and error, that is, through rigorous experimentation. Learned medical debates were conducted by exalting the authors' own experimental authority and questioning their opponents' reliability and experimentation. The baroque language of metaphors and emblems lost its appeal. The very different writings on geology of Francisco Xavier de Orrio, a mid-eighteenth-century Spanish Jesuit teaching at Zacatecas, and Andrés Ibarra Salazán, a student at Mexico City's College of Mining at the turn of the nineteenth century, exemplify this cultural change. Whereas for Orrio, the Earth was a recently created organic macrocosm, in which mercury was transmuted into gold and in which hidden sympathies made stones resemble flora and fauna, for

Ibarra Salazán, the Earth had a long history, revealed in layers of rock and fossils that documented slow processes of geological change.[52]

Travelers and Cultural Change

The remarkably different fates of two foreign scientific expeditions in Spanish America speak volumes about the profound changes in cultural outlook that the Spanish American elites underwent in the second half of the eighteenth century. From 1735 to 1745, an expedition headed by three French academicians—Pierre Bouguer, Louis Godin, and Charles-Marie de La Condamine—stayed in the Andes in order to measure the length of a degree of meridian at the equator, allegedly in order to settle the Newtonian-Cartesian debate over the shape of the globe. The much-publicized expedition proved an unmitigated disaster for the French, arriving back in Paris eight years too late to have any impact on the final resolution of the debate.[53] Seeking to do away with charges of scholarly incompetence, La Condamine wrote an account of the travails of the expedition that is both painful and hilarious; more important, it shows that the French worked amid a hostile population. The expedition was not only plagued by accidents (such as instruments lost to strange meteorological phenomena and scientists assailed and killed by debilitating "tropical" fevers) but also doomed, for from the start the locals greeted it with open hostility. The Amerindians, portrayed by La Condamine as submissive and stupid, either destroyed, stole, or moved the markers for trigonometric calculations erected by the French on mountain summits. The natives also systematically refused to work as scouts for the Europeans, and when they agreed to become guides, they often fled and left the academicians stranded in the most rugged terrain. Amerindian porters twice managed to lose La Condamine's luggage. Blacks and *castas* were no more sympathetic. In the eyes of the French, they were unruly "plebes" who, in open defiance of European decorum, carried swords, with which one of them stabbed a servant of the expedition. Neither did the white Spanish elite shield the French from the hostility of the Amerindians, blacks, and mestizo plebeians. The academicians often faced the rage of both imperial and provincial white authorities, although warmly welcomed by "enlightened" Jesuits and by a select handful of scholarly Creoles (fig. 3.3). For years the French had to fight in court charges of engaging in illegal trade and of building unauthorized commemorative monuments, which bore self-congratulatory inscriptions that failed to mention the Spaniards. Finally, in the wake of a popular riot led by the provincial Creole elites of Cuenca,

FIG. 3.3. Frontispiece of a doctoral thesis defense published in
Quito and dedicated to the French academicians Bouguer, La
Condamine, and Godin at a Jesuit college in Quito, Ecuador, in
June 1742. Reproduced in Charles-Marie de La Condamine, *Journal
du voyage fait par ordre du roi à l'equateur* (Paris: Imprimerie Royale,
1751). The putti busy measuring and gathering "matters of fact" with
the help of all sorts of experimental apparatus may well signify a
turning point in Creole sensibility. Discourses on experimental and
mechanical philosophy arrived in the colonies in the second half of
the eighteenth century, along with new secular learned institutions.
Courtesy of the John Carter Brown Library, Brown University,
Providence, R.I.

in which the surgeon of the expedition, Jean Seniergues, was stoned and stabbed to death, after being accused of promiscuity and deflowering a local beauty (and in which the rest of academicians were forced to flee for their lives), the French got involved in a trial of the Creole ringleaders that lasted three years but led to no punishments.[54]

Some sixty years later (1799–1804), Alexander von Humboldt and Aimée Bonpland visited several Spanish American colonies. Throughout their journey, these two traveling scientists were greeted as heroes. Upon their return to Paris, Humboldt published thirty volumes of observations and philosophical reflections, which, unlike La Condamine's, portrayed the Spanish American colonies most favorably. The imperial authorities and local literati not only embraced the Europeans warmly but, more important, gave the foreigners the results of forty years' worth of their own collective investigations. Humboldt's thirty volumes should be read not only as the product of a genius working in isolation but also as a summary of the Spanish American Enlightenment.[55]

A Unifying Theme

Changes in Creole cultural sensibilities were also reflected in the new scientific idioms they chose to express their age-old patriotic longings. Thomas F. Glick has shown that although the new scientific institutions created by the Bourbon regime in the colonies were, by and large, staffed and led by peninsulars, they contributed to the training of cadres of patriotic Creole scientists, who spearheaded the wars of independence against Spain (1810–24), having become aware of their status as colonized subjects much earlier than other sectors of the population. According to Glick, Creole scientists became "Newtonian" liberals who sought to create national sciences around the defense of Andean and Nahua taxonomies (threatened by the expansion of the new Linnean botanical classifications, which arrived in the colonies along with tactless Spanish imperial scientists), the identification and development of local materia medica, distinct from European ones, and resistance to negative European characterizations of the American climate.[56]

Creoles modified the scientific idioms in which they cast their proto-nationalism. Whereas baroque patriots had used astronomy and astrology to exalt God's providential designs and had praised the mineral and pharmaceutical wonders of their land, late eighteenth-century scholars sought to tap the agricultural potential of the colonies. They argued that each colony was endowed by Providence so that it would become a leading commercial emporium in the world. Naturalists presented their local territories as micro-cosms of the globe in which the multitude of ecological niches and endless equatorial agricultural cycles made the land capable of supplying every need of the world's markets. These naturalists also assumed that the natural laws of the Americas were different from those of Europe, and that New World phenomena could be studied only by Creole scientists. José Antonio de Alzate, a leading Mexican naturalist and editor of several periodicals, insisted that Mexico's rare natural productions undermined and upset all sci-entific hypotheses devised by Europeans and sought to create a science that only Mexicans could foster and interpret.[57] Peruvian Creole physicians took advantage of humoral theory to create a form of medical nationalism that maintained that the climate, bodies, and diseases in Peru were singular, and thus that only Peruvian physicians could identify and cure local diseases.[58]

To be sure, profound cultural differences separated the worlds of baroque and "Newtonian" Creole scientists. But patriotism remained a constant uni-fying theme throughout the long eighteenth century.

New World, New Stars
Patriotic Astrology and the Invention of Amerindian and Creole Bodies in Colonial Spanish America, 1600–1650

In 1646, angered by the attempts of certain sectors of the Spanish clergy to bar him from a recent appointment as procurator general in Rome of the Franciscan province of New Spain, Buenaventura Salinas de Córdoba (d. 1653) wrote a memorial from Italy to King Philip IV of Spain requesting support. Facing the criticism of his opponents, who assumed that the New World had turned Spanish American colonists into degenerate Amerindians, Salinas de Córdoba portrayed America as a biblical paradise. The benign astral influences of the New World were demonstrated not only in the fact that the Indies were a veritable microcosm in which everything in the world could be found, but, more important, in the character of its peoples. Amerindians, Salinas de Córdoba averred, were gentle, meek, and generous, ideal material for creating communities of saints all over America. Spaniards born in the Indies, also known as Creoles, had outstanding minds, which, once trained in colonial universities, entitled them (by divine and natural right) to occupy the highest bureaucratic posts in the land.[1] Yet, for all his praise of Amerindians and Creoles, Salinas de Córdoba also clearly differentiated among those living in America. He reminded the king that Creoles were the heirs "not of the Indians" but "of the valor and blood of Spain."[2] In fact, in another memorial to the king, published in Lima in 1630, Salinas de Córdoba had argued that skin color and behavioral differences among nations could not be explained by environmental factors. Taking issue with many "learned authors" (*varones doctos*) who had maintained that the physical features of Amerindians and blacks stemmed from the temperament of the land and/or heavenly influences (for example, warmth and sunny climates produced black skins), Salinas de Córdoba insisted that Amerindians and blacks owed their color and servile demeanor to Noah's curse on the descendants of Ham.[3]

There is, to be sure, no novelty in maintaining that European colonists sought to differentiate themselves racially from Amerindians. Spanish America was, after all, a society built on corporate privileges and social estates, which, moreover, overlapped with racial hierarchies: although *castas* (mixed bloods) occupied the interstices of the original three-tier system of Spaniards, Amerindians, and Africans, blurring the colonial boundaries of class and race, Spanish America was a society obsessed with identifying and enforcing racial hierarchies.[4] But Salinas de Córdoba's treatises reveal much more than a quaint exercise in white European self-identification. About the time Salinas wrote his memorial of 1646 to the king, Antonio León Pinelo (1590–1660) was seeking to prove that paradise lay hidden somewhere on the eastern slopes of the Peruvian Andes. After a successful career as a magistrate in Lima, the expatriate León Pinelo was recalled to serve on the Council of the Indies in Madrid, where he set out to defend America from Europe's negative characterizations. León Pinelo maintained that Peru was so bountiful and temperate that it was the site of the Garden of Eden, and that the Amazon, Magdalena, La Plata, and Orinoco were the four rivers of paradise mentioned in Genesis. To prove that the Amazon was the biblical river Gehon that ran through the land of Ethiopia, León Pinelo insisted that the torrid zone of the Indies and Africa were one and the same. To clinch his argument, he argued that the Amerindians were like the Ethiopians, "black," slavish, cursed descendants of Ham.[5]

León Pinelo was well aware of the contradictory nature of his argument. How could the idle, meek, "black," cursed Amerindians be native to the Garden of Eden? Judged by the quality of its people, Peru clearly was not paradise. The Italian humanist Julius Caesar Scaliger (1484–1558) contended that paradise must be somewhere in Europe, to judge by the merit and intelligence of its people, but León Pinelo insisted that its location should be determined solely by the quality of the land, not peoples: nations moved around (which made it impossible to determine whether the Amerindians were truly the original inhabitants of America), and they also had set mental and physical differences acquired from creation that could not be changed by astral or climatic influences.[6] Like Salinas de Córdoba, León Pinelo introduced the notion that the environment only slightly affected the character of peoples, which otherwise was innate. Both authors posited that the bodies of European and Amerindians were radically different.

In this chapter, I argue that Salinas de Córdoba and León Pinelo belong to a larger current of thought among learned seventeenth-century colonists, who, as they sought to defend America from negative European characterizations, invented modern forms of the racialized body that scholars have wrongly attributed to the rise of modern science in Europe in the eighteenth and nineteenth centuries. I contend that the science of race, with its emphasis on biological determinism, its focus on the body as the locus of behavioral-cultural variations, and its obsession with creating homogenizing and essentializing categories, was first articulated in colonial Spanish America in the seventeenth century, not in nineteenth-century Europe. Using the European sciences of the day, astrology and Hippocratic-Galenic physiology, learned colonists articulated a form of the "science" of race that claimed that there were innate bodily and mental differences separating peoples from one another. They maintained that although constellations and climate could in fact render white colonists more intelligent than the Europeans, bodies remained impervious to environmental and cultural influences: the climate and stars of America could not transform white Europeans into Amerindians, any more than the cold air of Europe could turn a black person white.

New World intellectuals showed little interest in theories of "generation," or heredity, that denied Amerindians and Europeans a common Adamic ancestry, which potentially might have made their argument that there were innate racial differences heretical.[7] They never challenged the authority of the Bible and avoided the contemporaneous rancorous debates on the subject in Europe, where Isaac de La Peyrère (1596–1676) and the so-called Pre-Adamites maintained that most nations had originated in separate creations.[8] Rather than attributing racial differences to polygenism, Creole scholars held that the different physical and mental traits of Amerindians and blacks were the result of Noah's curse on Ham.[9]

In a thesis now widely accepted, John H. Elliott has argued that what was "new" about America was easily incorporated into ancient classical and biblical paradigms during the early modern period and therefore had only a blunted impact on European consciousness.[10] The story I am about to tell both confirms and undermines Elliott's thesis. Spanish American intellectuals held that the body was immune to environmental influences and therefore went against ancient Mediterranean paradigms of behavior and skin color being determined by the close interactions of the individual with the stars and the climate. But this challenge to ancient views, inadvertently developed by colonial intellectuals, did not influence, and was not even acknowledged in, later European discourses of the racialized body. Postcolonial theorists have called for a more nuanced and global understanding of the history of processes and institutions that to this day have been narrowly construed as European.[11] Ann Laura Stoler, for example, has used such postcolonial approaches to shed light on the deep colonial roots of nineteenth-century Victorian sexuality and has offered a new, more provocative reading of Michel Foucault's Eurocentric *History of Sexuality.*

Stoler forcefully maintains that the colonies were "laboratories of modernity," where "key symbols of modern western societies [such as] liberalism, nationalism, state citizenship, culture, and Europeanness were first clarified among Europe's colonial exiles and by those colonized classes caught in their pedagogic net in Asia, Africa and Latin America, and only then brought home."[12] This chapter shows that the seventeenth-century Spanish American colonies were indeed laboratories of modernity. But my story, unlike Stoler's, cannot be used to demonstrate continuities in the colonial roots of European modernity, because none of the ideas created in Spanish America later proved influential in Europe.

Precisely at the time when the European core was articulating a racialized modern view of the body, a new rhetoric of rigor and objectivity in science declared the ideas on the subject that had once been created on Europe's colonial peripheries to be useless and unreliable.[13] It is not surprising, therefore, that the sixteenth- and seventeenth-century origins of modern views of the racialized body have long been forgotten. To rediscover these deep colonial roots, we need to restore some of the credibility that early modern science once enjoyed.[14] The science used by Spanish American intellectuals came wrapped in the ancient Mediterranean idioms of Hippocratic-Galenic physiology and astrology. Astrology was considered a very serious science, which studied the processes through which planets and fixed stars controlled "generation and corruption" in the sublunar world by eliciting change among the four elements (water, earth, fire, and air) and therefore over human temperaments and constitutions, that is, over the bodily balance of elements and humors as described in the Hippocratic and Galenic corpus. Astrology was part of the obvious mental landscape of every learned individual in the early modern world, regardless of religion or country of origin. Although prognostication was itself a contentious issue that raised all sorts of theological and political questions, everybody took it for granted that the stars affected behavior in the sublunar world.[15]

That Hippocratic-Galenic physiology and astrology helped Europeans make sense of their puzzling encounter with the New World has not been sufficiently recognized. Several generations ago, Antonello Gerbi studied the patriotic responses triggered in the eighteenth and nineteenth centuries in the Americas, including the United States, by the belief that the New World was a degenerating tropical environment. Little is known, however, about European characterizations of the Indies in the sixteenth and seventeenth centuries. By the late sixteenth century, America had come to be perceived as a degenerating land; it was thought to be overly "humid" and thus emasculating, but, more important, it was assumed to be ruled by those new, negative constellations that Europeans had recently discovered and charted in the Southern Hemisphere. Such representations prompted Creole and émigré European scholars to react, creating forms of patriotic astrology. A full-fledged, consistent defense of the environment, however, could only undermine tenets most settlers assumed to be true, namely, that Amerindians were slow-witted phlegmatics who had to be forced to work. How to defend a land from European innuendos that had created nations of inferior, mentally challenged Amerindians? I argue that the reaction of the colonists was to postulate clear-cut racial distinctions, and that Creoles and Amerindians (and, to a lesser degree, blacks) had different kinds of bodies.

In the first section of this chapter, I explore Europe's negative climatological and astrological characterizations of the New World, as well as the views of the European body in the Indies these views engendered. In the second section, I study the reaction of colonists in Spanish America, including the rise of a genre of patriotic astrology and the theory that white European colonists and Amerindians had different types of bodies. In the third and final section, I discuss the historiographical import of understanding these views.

Discovering America's New Constellations

From the early sixteenth century on, learned Europeans felt the need to explain why a land that ancient cosmographers had predicted to be uninhabited was in fact temperate. Gonzalo Fernández de Oviedo (1478–1557) was perhaps the first European to offer, in 1526, a list of meteorological mechanisms to explain why Spanish America, a region largely located in the torrid zone, was temperate and bountiful and not barren, despite the scorching equatorial sun. Oviedo suggested several cooling mechanisms: days and nights had the same length year round in the tropics, which thus had longer nights to cool off in summer than Europe; high mountain ranges and vast oceans encircling an elongated continent contributed to cooling the land; and large rivers and wet soils dampened the heat.[16] As conquistadors discovered large river basins, lakes, and tropical forests, a sense that America was a temperate, yet humid, continent came to dominate in the imaginations of European scholars.[17] In 1555, Girolamo Cardano (1501–76), a distinguished Italian humanist, argued that climate was determined by the height of land relative to the sea. The torrid zone was temperate because it lay below sea level and attracted water from the poles, which Cardano thought to be the globe's highest continental masses. This polar-equatorial flow of water, Cardano maintained, made tropical America temperate and moist.[18] In 1589, the Jesuit José de Acosta (1540–1600) offered the first detailed account of the meteorological mechanisms that kept tropical America temperate and noted that the New World was humid, although the closest thing on earth to paradise.[19] The abundance of water made large sections of America uninhabitable; humidity was also partially responsible for the numerous earthquakes there.[20]

Oviedo, Cardano, and Acosta did not draw any negative conclusions from the humidity of the New World. However, in 1579, the Franciscan Diego Valadés (b. 1533) asserted in Italy that the Amerindians were "stupid because they are born in thick air."[21] In 1591, Juan de Cárdenas (1563–1609), an émigré Spanish physician long resident in the New World, maintained that humidity in the Indies was not only the cause of frequent earthquakes (for vapors were trapped in America's cavernous subsoil) but also sapped the strength of the population (humidity caused chronic illnesses that weakened, dried, and consumed the human body).[22] By the early seventeenth century, it had become a truism that the humidity of the Indies debilitated organisms, made roots shallow, and caused food to lack nourishing value.[23] Colonial physicians in Mexico told the English apostate Thomas Gage (1603–56), who traveled in New Spain in the early seventeenth century, that the reason he always felt hungry in America even after having huge meals was because the humidity of the land made meat attractive, yet short in "substance and nourishment," fruit delicious, but with "little inward virtue," and the people, like the meat and the fruit, good-looking, but "false and hollow hearted" inside.[24]

The purported humidity of America was a claim with heavy ideological baggage. Scholarship had associated masculinity with warm, dry environments, and femininity with moist, cold ones. America posed the threat of impending sexual transformations for colonists.[25] Some Spanish authors contended in the course of the sixteenth century that America was a land where women urinated standing, while men did so seated.[26] But the threat of emasculation associated with humid environments was offset by the contention that tropical America was bountiful and paradisiacal because compensating meteorological mechanisms made it temperate.[27]

More challenging for colonial intellectuals were the writings of Europeans on the negative effects of the stars of the New World. The notion that the stars did not merely influence the body but gave societies their unique and distinct characteristics was an ancient one, widely influential in early mod-

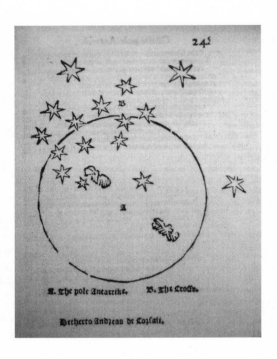

FIG. 4.1. Star chart of the Southern Hemisphere by Andrea Corsali (1517), including the Southern Cross and two Magellanic Clouds. From Pietro Martire d'Anghiera, *The decades of the newe worlde or west India*, trans. Richard Eden (London, 1555). Courtesy of the John Carter Brown Library, Brown University, Providence, R.I.

ern Europe.[28] Despite the fact that the Portuguese had slowly made their way down the coast of Africa through the fifteenth century, European audiences had no knowledge of most stars south of the equator up until 1502, when Amerigo Vespucci (1451–1512) offered the first drawings of southern constellations.[29] Vespucci introduced learned Europeans to some twenty new stars and marvelous new heavenly phenomena.[30] It is fair to argue that Vespucci assigned the name "New World" to the Americas largely because the skies there were populated by "stars and signs" unknown to the ancients. Vespucci was so fascinated by the heavenly marvels of the Southern Hemisphere that he evidently wrote a treatise on its constellations (which did not, to my knowledge, survive him). He thought the southern stars bigger and brighter than those over Europe and connected them with the mechanisms that made tropical America so wonderfully temperate.[31] Confirming his view of the New World as a paradise enjoying benign astral influences, Vespucci found the land to be inhabited by peoples who, although they ate each other and went naked without shame, were physically perfect and enjoyed extraordinary longevity, without wrinkles, sagging breasts, or stretch marks.[32] In 1517, Andrea Corsali, an Italian navigator who sailed to the East Indies under the flag of the king of Portugal, published a letter addressed to the Medici,

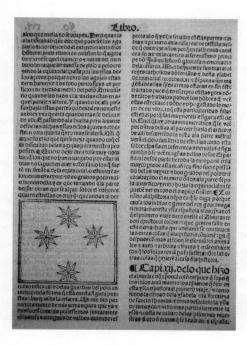

FIG. 4.2. Representation of the Southern Cross in Gonzalo Fernández de Oviedo, *La historia general de las Indias* (Seville, 1535). Courtesy of the John Carter Brown Library, Brown University, Providence, R.I.

with a map of the stars of the Southern Hemisphere (fig. 4.1), in which he waxed eloquent about the beauty of the Southern Cross, so bright "that no other heavenly sign may be compared to it."[33]

Fernández de Oviedo was the first learned European to cast doubts on such positive characterizations of the new constellations. In 1526, Oviedo argued that even though they were uncharacteristically slow, the animals in America that looked like tigers were, in fact, tigers, and that their phlegmatic character could be accounted for by the climate and constellations that ruled the land, the same stars that made its people "shy and cowards."[34] Oviedo's description of heavenly influences over America, however, was ambiguous. In 1535, in his much awaited *Natural History of the Indies*, Oviedo offered an illustration of the Southern Cross (fig. 4.2) and declared that those who, like himself, had lived under the influence of this constellation were condemned to toil and suffer. Yet he also presented the Cruzero as his own, for Charles V had made a representation of the Southern Cross the main emblem of the coat of arms granted to Oviedo for having written on the marvels and natural history of the New World (fig. 4.3). In the imaginations of the emperor and Oviedo, the metonym of the Southern Cross, a new constellation, stood for the whole of America.[35]

FIGURE 4.3. Gonzalo Fernández de Oviedo's coat of arms, which includes the Southern Cross, from his *La historia general de las Indias* (Seville, 1535). Courtesy of the John Carter Brown Library, Brown University, Providence, R.I.

Oviedo's misgivings about the effects of the constellations of the New World only fueled the curiosity of European scholars. In a selection of translations of Spanish and Italian sources that included the first three decades of Pietro Martire d'Anghiera's *De orbe novo*, published in 1555 under the title *The decades of the newe worlde or west India*, Richard Eden (1521–76) dwelt on the singularities of the land, sea, and, more important for the purposes of this chapter, skies of the New World. The long title page of the book whetted the appetite of potential buyers by warning them that "[in this book] the diligent reader may not only consider what commodities may hereby chance to the whole Christian world in time to come, but also learn many secretes touching the land, the sea, and the stars."[36] Eden, who was convinced that the "miraculous mouinges of the Planets, Starres, and heauens [caused . . .] the varietie of times and the dyversitie of all natural things . . . [particularly] the varietie of divers complexions, formes and dispostions of all creatures under the face of heauen," also believed that the New World was a singular land whose new stars merited as much attention as did descriptions of its already better known fauna and flora.[37] Eden's *Decades* contained illustrations of southern constellations and passages from Vespucci, Antonio Pigafetta (ca. 1480/91–ca. 1534), and Corsali, all of whom had agreed that the stars of the New World were beautiful, bright, and benign.[38] The book also included Oviedo's *Sumario*, however, with its negative comments on the degenerating effect of American constellations on "tigers" and people alike.[39]

Julius Caesar Scaliger, himself a renowned humanist and the father of Joseph Scaliger (1540–1609), one of the greatest polymaths ever to appear in Europe, may have contributed to consolidating the party of those who had misgivings about the effects of America's heavenly constellations. In a work written in 1557 aimed at debunking the views of Cardano on a variety of issues, Scaliger took issue with the theories of the Italian humanist on the origin of metals. Cardano had argued that the sun was responsible for the generation of gold and precious stones, and that an eastern equatorial position on the globe was therefore enough to make a land rich. Scaliger dismissed this proposition out of hand, because among other things "wealth" and "east" were relative categories. Europe, Scaliger argued, although it might indeed be poorer in gold, was in fact richer than "eastern" tropical lands, because it was rich in iron mines. It was also, unlike any other, a land rich in brilliant minds. Using Brazil as typifying tropical lands, Scaliger maintained that, for all its proximity to the sun, Brazil was a land without gold and precious stones, a land of spices that were "scarce, ignoble, and bad," and a place where the way of life was wild and bestial.[40] Clearly, neither the sun nor the constellations of the New World had done much to help this land and its peoples.

Scaliger's misgivings about America's astral influences were part of a larger scholarly environment. In the second half of the sixteenth century, the first extended analyses of the corrupting effects of America's stars and heavenly signs were penned in Mexico. The Franciscan friar Bernardino Sahagún (d. 1590), whose encyclopedic studies helped preserve Nahua lore, believed that the demographic, moral, and economic crisis then wreaking havoc in Mexico, which shattered Franciscan hopes for a millenarian Church in the Indies, had been caused by evil astral influences. Sahagún argued that the Franciscan failure to create an indigenous priesthood at the College of Santa Cruz at Tlatelolco typified the origins of Mexico's larger national crisis. The Franciscans, unable to understand that heavenly constellations had caused Amerindians to be lascivious and idle, and that only strict corporal and labor discipline could offset such negative astral influences, had relaxed their original high monastic standards, causing Amerindians to slip back into promiscuity and idleness. According to Sahagún, Aztec lords had long understood the dangers of the constellations and had therefore punished drunkards, vagrants, adulterers, and liars harshly. Only a reversion to traditional high standards of Aztec discipline, Sahagún suggested, could save Santa Cruz and Mexico from total collapse. To clinch his argument, Sahagún presented the Creoles as examples of the malignant and degenerating effects that the stars of the New World had over people living in it.[41] Francisco Hernández was sent to the Indies by Philip II in the 1570s solely to identify medicinal plants to enrich the royal pharmacy but nevertheless also wrote a book on Mexican antiquities, because he thought that human conduct, just as much as the virtue of plants, was conditioned by the stars.[42] Like Sahagún, Hernández believed that the stars limited free will, and he explored the negative effects that the constellations of the New World had on Amerindians. He found them idle, docile, untrustworthy, and obedient only to force, and therefore to be disciplined according to ancient Aztec lore through harsh regimented labor. Hernández worried that Creoles "obedient to astral influences" were adopting "the [idle] habits of the Indians."[43]

Sahagún's and Hernández's views were shared by many and were part of a larger scholarly mood concerned with the debilitating effects of the new constellations and heavenly phenomena of America.[44] In 1590, José de Acosta openly denounced those who had argued that the stars of the New World were brighter and more numerous than those in Europe. He found the stars of America to be in fact smaller and much dimmer. Although Acosta praised the rest of nature in America, the only positive thing he was able to say about heavenly phenomena in the Indies was that the skies were "singular," for numerous dark holes in the Milky Way unobservable from Europe could be spotted, adding therefore to his message that the skies of America were lacking in stars and brightness.[45] In a passage destined to become canonical—largely because Antonio de Herrera y Tordesillas (d. 1625), the royal chronicler of the Indies and the author of the influential *Historia de los hechos de los castellanos en las Indias*, copied it verbatim in 1601[46]—the Italian humanist Giovanni Botero (1540–1617), like Acosta, concluded in 1596 that the stars and constellations of America were "inferior." Botero readily drew very negative conclusions from this "fact" (along with others such as that the Indies lacked large quadrupeds and valuable fruits and spices and that the continent's topography and overall shape hampered navigation, overland trade, and commerce). Europe, Botero concluded, was "better suited for human life." It was not only superior to America in quality of land, material prosperity, and civilization but also was under more benign heavenly influences.[47] The stars of the New World made native Americans idlers and drunkards, who had to be disciplined by means of regimented labor.[48] In 1612, Juán de la Puente, an apologist for Habsburg claims to universal monarchy, argued that Spaniards, who had once colonized all the ancient world and who had recently established colonies in Asia, Africa, and America, had to acknowledge that the heavenly constellations of America had sapped their courage and roving spirit. Although rich in metals and plant life, de la Puente contended, the Indies produced only degenerate humans.[49] "The heavens of America induce inconstancy, lasciviousness and lies," he argued, "vices characteristic of the Indians and which the constellations make characteristic of the Spaniards who are born and bred there."[50] In 1617, Samuel Purchas (1577–1626), the English editor of an influential compilation of travel accounts, drawing on Acosta and Botero, concluded that the new heavenly constellations of America had smaller and dimmer stars, that the sun stayed longer over Europe than over the Indies in the course of a year, and that the "want of sun and stars" had rendered the New World colder, with fewer animals, spices, and fruits, and, of course, with little to show in way of intellectual life.[51] Even some Europeans who supported the interests and aspirations of the Creoles accepted this type of negative characterization of the American stars. For example, León Pinelo, the magistrate who contended that the original paradise had been located in Peru, depicted the stars of New World as less numerous in the frontispiece to his *Epitome de la biblioteca oriental y occidental* (1629), which shows a woman representing the East Indies crowned by fourteen stars, while a female figure representing America has only twelve stars over her head (fig. 4.4).[52]

In the seventeenth century, such European negative views of the stars and climate of the New World biased perceptions of the bodies of Amerindians and Spaniards living in Spanish America. Most European scholars assumed that Amerindians and Europeans were descendants of Adam and therefore had similar bodies and minds. These essential similarities, however, were thought to have been modified by the climate and stars with "accidental" variations. For, in the words of one of the first English translators and editors of Iberian and Italian travelers' tales, Richard Eden, "the complexion and strength of body of [all the inhabitants of the world] are proportionate to the climate assigned to them, be it hotte or colde."[53] The seventeenth-century literature on the origins of Amerindians reveals how the understanding of America as a humid continent with negative heavenly constellations helped scholars explain the transformation of white bodies into Amerindian ones without having to postulate two separate racial types.

The Spanish Dominican friar Gregorio García (d. 1627) is typical of the learned Europeans who looked to environmental influences to explain the

Fig. 4.4. Frontispiece to Antonio de León Pinelo's *Epítome de la biblioteca oriental y occidental* (Madrid, 1629). Note that Oriens has fourteen stars, whereas Occidens has only twelve.

origins of a characteristic Amerindian body type. In 1606, in an encyclopedic book seeking to explain the origins of the Amerindians, who, he contended, had come to the New World from Asia, Africa, and Europe, García
maintained that the natives might have descended from the Carthaginians.
García was hard pressed, however, to explain how courageous North African
Carthaginians had become hairless, cowardly, pusillanimous Amerindians.
García argued that the humidity and the constellations of tropical America
had turned the Carthaginians into effeminate, cold, and humid Amerindians.
The humid constitutions of the Amerindians, according to García, explained
why they, like women and *castrati*, did not have beards and were passive,
stupid, and slothful.[54] The stars of America in particular were responsible
for the major organic transformations witnessed in the continent. García
considered that many animals, such as the buffalo and llamas, were not only
unique to the continent but also "monsters." These animals had developed
monstrous bodily features away from their original Old World parents (the
bull and the camel) due to the continuous astral and climatic influences of
the land. These differences had become fixed, in much the same way that
the climate and stars of the continent had permanently emasculated and
darkened the former Carthaginians.[55] The degenerative dangers for contemporary white Spanish colonists implicit in this account were not lost on
García, who, however, sought to tone down the consequences of his theory
by claiming that Creoles and long-term European residents in the Indies
were better prepared. Unlike the Carthaginians, who had eaten American
roots and nonnutritive fruits, they were eating only nourishing European
staples.[56]

García's views on the Amerindians and heredity lingered on. In a work
written in Yucatan but published in 1633 in Valladolid, Spain, the Franciscan
friar Bernardo de Lizana (1581–1631), astonished by the beauty and grandeur of Maya ancient buildings and the complexity and extension of their
polities, insisted that the Maya were the descendants of Carthaginians, for
only a nation like the Carthaginians could have had the intellectual skills
required to design such buildings and the military valor to conquer and
consolidate such empires. Yet Lizana, who thought that the contemporary
Maya were childish and brutish, argued that climate and total isolation from
Carthage had been responsible for transforming ancient Carthaginians into
barbarous and crude (*toscos*) Maya.[57] The views of García and Lizana were
even echoed by some Creole authors. In 1681, Diego Andrés Rocha published a book in Lima on the origins of Amerindians in which he sought to
emulate the baroque erudition of Gregorio García. Rocha thought that the

Amerindians were the descendants, not of Carthaginians, however, but of the primitive inhabitants of Spain, although the Amerindians were cowards and lacked beards. Amerindians and Spaniards possessed the same basic bodily constitution, he contended, to which new "accidental" features had been added by the American constellations and climate. Like García, Rocha argued that the ancient Spaniards-turned-Indians had become cold and humid and had therefore lost their martial prowess and beards. Like García, Rocha also argued that Creoles had somehow been spared from rapidly becoming effeminate, stupid Amerindians thanks, in part, to the European food they ate and to the fresh influx of European blood that arrived in the colonies with every generation (which did not allow the colonists to cut all racial ties with their homeland); these two processes had slowed and even stopped the environmentally induced degeneration.[58]

Patriotic Creole Astrology

Creole scholars reacted angrily to the negative characterizations of the stars and constellations of the New World. It was precisely at the time when such negative characterizations had come to dominate Europe's perception of the heavenly constellations of the New World that a Creole consciousness began to emerge in Spanish America.[59] At the turn of the seventeenth century, Creoles began to flood the court in Madrid with memorials pleading for an extension on grants to all existing *encomiendas* (grants in indigenous labor and tribute given by the Crown to the leading conquistadors). Conquistadors' descendants complained that Spain was reneging on its commitment to foster a class of New World grandees, privileged landed nobilities wealthy enough to care for their Amerindian retainers and underwrite the welfare and defense of the new viceroyalties. Creoles articulated a somewhat misleading view of themselves as dispossessed nobles outcompeted by ravenous, transient, peninsular upstarts. Also at the same time, disputes broke out between peninsulars and Creoles inside religious communities over which group had the right to govern them. The *alternativa*, an arrangement devised during those heady years to rotate the government of religious orders among Creole and peninsular friars, only made things worse by reifying and essentializing identities and thus heightening conflict.[60]

It is not surprising that in this context, colonial intellectuals began to write against negative European representations of the Americas. The second quarter of the seventeenth century witnessed the maturation of a genre of patriotic astrology in which the heavenly influences of America were consistently cast as having soothing and beneficial effects, revealing God's providential design. Astrology spread into every corner of colonial culture. Studies by Irving Leonard and Elías Trabulse tracking the inquisitorial trials of astrologers suggest that astrology was little tolerated in Catholic countries, particularly after the Council of Trent, because it was seen as denying freedom of will, and hence as almost heretical.[61] But this is deeply misleading. Although the Inquisition often targeted those who sought to communicate with evil astral intelligences (i.e., demons) or who indulged in prophecy and political prognostication, it actually forced those who cast doubt on the scientific status of astrology to recant.[62] As the science that studied the influence of stars on temperaments and constitutions, astrology was in fact officially sanctioned: colonial universities had chairs of medical astrology, and most rituals of power were thoroughly permeated with astrological metaphors.[63] It is no wonder therefore that astrology was openly and defiantly deployed in self-defense by the colonists.

The Augustinian friar Antonio de la Calancha was perhaps the first Creole writer to pen a sustained astrological defense of America, and he found that the "stars and signs" of the New World revealed God's special design for it. In a treatise published in Barcelona in 1638, Calancha alluded to two separate unpublished volumes he had written on the constellations of the Southern Hemisphere. These studies, he maintained, would supersede anything Europeans had previously written on the subject. In the heavens of Peru, he averred, he had found "a bounty [*manos llenas*] of heavenly wonders, and whole pieces of sky crowded with stars unknown to astrologers and sailors."[64] According to Calancha, the stars of America were the largest, brightest, and most numerous in the world; in fact, most of the forty-eight constellations that Europeans had identified in the skies lay in the Southern Hemisphere. Short excerpts from astrological writings of Calancha's giving the horoscopes of all the major Spanish American cities and provinces (which do not otherwise appear to have survived him) found their way into his 1638 treatise, identifying the ascendant stars of a handful of important towns in the viceroyalty of Peru, which he found without exception to be under most benign astral influences.[65] Calancha also maintained that heavenly phenomena that were portentous in Europe or the rest of the world would lose their potency in the Indies. Calancha thought, in fact, that Peru was blessed; God had chosen to protect it by giving it not only crosslike constellations but also crosslike fossils, stones, and plants.[66]

In 1646, in a natural and moral history written in part to prove that Chile was one of the most temperate and bountiful lands in the world, the Creole

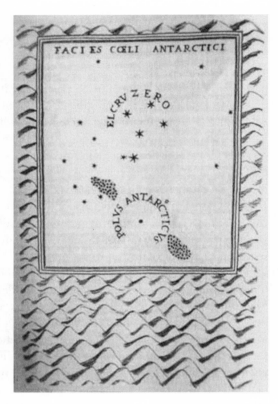

Inside the image: FACIES CŒLI ANTARCTICI · EL CRVZERO · POLVS ANTARCTICVS

FIG. 4.5. "Face of the Antarctic skies" in Alonso de Ovalle's *Histórica relación del reyno de Chile* (Rome, 1646). This chart is a variation of Corsali's (fig. 4.1). Like Corsali's, it includes the Southern Cross and two Magellanic Clouds.

Jesuit Antonio de Ovalle (1601–51) found the skies of Chile exceptionally good and contrasted the misty opaque skies of tropical America with the clear, bright ones of the Southern Cone (fig. 4.5). For his brand of patriotic astrology, Ovalle drew upon the studies of a Dutch disciple of the distinguished Flemish scholar Petrus Plancius (Pieter Platevoet [1552–1622]), Pieter Dircksz Keyser, who, along with Frederick de Houtman (1570–1627), produced the most complete catalogue of southern stars then available, shortly before dying aboard a vessel bound for Madagascar and the East Indies in 1597 (fig. 4.6). Keyser's catalogue was later used by European cosmographers and astronomers such as Plancius, Willem Janszoon Blaeu (1571–1638), Johann Bayer (fl. 1600), Jacob Bartsch (1600?–1633), Jodocus Hondius (1563–1612), and Johannes Kepler (1571–1630). Referring to Keyser as a "learned astrologer and pilot," Ovalle listed fourteen new constellations: Chameleon; Aspid Indica (New World Serpent); Volans (Flying Fish); Dorado (Swordfish); Hydra (Water Snake); Toucana (Toucan); Phoenix;

FIG. 4.6. A typical seventeenth-century sky chart of the Southern Hemisphere, which includes Dircksz and Houtman's new southern constellations. From John Seller, *Atlas Maritimus* (London, 1675). Courtesy of the John Carter Brown Library, Brown University, Providence, R.I.

Grus (Crane); Columba (Noah's Dove); Indus (Indian); Pavo (Peacock); Apus (Bird of Paradise); Triangulum Australe (Southern Triangle); and Crux or Cruzero (the Southern Cross). Ovalle overcounted the stars given in Keyser's and Houtman's original catalogues and concluded that the stars over Spanish America were more numerous, larger, brighter, and more benign than those in Europe.[67]

A third strategy of Creole patriotic astrology was that pursued by the seminal Mercedarian Creole natural philosopher Diego Rodríguez, the first holder of the chair of mathematics at the University of Mexico and author of highly sophisticated mathematical treatises. Rodríguez maintained that the heavens of New Spain were under the protection of the Immaculate Conception, for in Mexico "there is no [heavenly] sign that, although shocking and surprising for the ignorant, does not serve the Queen of Heaven and [help] explain her glories."[68] In clarification, and echoing a broader intellectual movement among Counter-Reformation intellectuals that sought

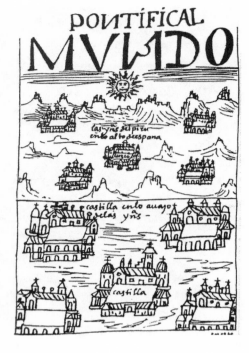

FIG. 4.7. The "Indies of Peru" lie
closer to the sun than Castile, which
is "below" the Andes. Proximity to the
benign astrological influence of the sun
make the Indies better than Spain. From
Guamán Poma de Ayala, *Nueva crónica y
buen gobierno* ([ca. 1615] Madrid, 1985).

to depaganize the heavens by means of alternative Christian readings,
Rodríguez charted the path of the comet of 1652 from the constellation of
Columba (Noah's Dove) to that of Medusa in order to show that it was symbolically associated with the Immaculate Conception. Columba, Rodríguez
explained, stood for purity, like that of Mary, whose immaculate conception
had spared her a postlapsarian human nature, while Medusa represented the
dragon that sought to kill the pregnant Virgin's child in Revelation 12:3–5.
Since the image of the Immaculate Conception was a Virgin eclipsing the
sun, Rodríguez also asserted that in a Mexico under the protection of Our
Lady of Guadalupe, eclipses, or any other heavenly phenomena (including
the comet of 1652), could only be harbingers of joyous news. For example,
the path of the comet through Mars, which symbolically stood for wisdom,
showed that it would make the colonial authorities wise. The fantastic battles
of Our Lady of Guadalupe in Mexico against the dragon of Aztec idolatry
were thus recapitulated in the skies of Mexico, proclaiming its safety and
merits.[69] The genre of patriotic astrology proved so appealing that even a
member of the acculturated Amerindian intelligentsia deployed it for his

own purposes. Felipe Guamán Poma de Ayala (fl. 1613), an early-seventeenth-century Amerindian chronicler from the province of Huamanga, Peru, argued both in painting and in prose that the Andes were closer to the sun than Castile, and therefore were richer and better, for, according to "philosophers, astrologers and poets," the sun had important positive astrological influences over countries (fig. 4.7).[70]

Amerindian Bodies

Patriotic astrology notwithstanding, Creoles faced a seemingly insurmountable challenge when they sought to defend the reputation of the New World. To sustain the notion that the New World was a temperate paradise implied, of necessity, arguing that the Amerindians might be superior to Europeans, because only uniquely gifted people could possibly be born in the original Garden of Eden. Bartolomé de las Casas (1474–1566), the spirited Spanish Dominican who devoted his life to the protection of Amerindians, showed the dangers of a sustained logical defense of the American environment when he demonstrated in the 1550s that praising the New World for its paradisiacal qualities implied the mental and physical superiority of Amerindians over Europeans. After waxing eloquent about the beauty of the land and describing the mechanisms that made the torrid zone temperate in the New World, Las Casas maintained that America was one of the most salubrious environments on earth. Consistent with this hypothesis, he argued that Amerindians were exceptional human beings. He drew on traditional Hippocratic-Galenic medical physiology and medieval faculty psychology to argue that since the natives lived in extremely temperate climates, they therefore had exceptionally good mechanisms of perception and superior intelligence. Moreover, Las Casas argued that, in addition to the American environment contributing to producing Amerindians whose five internal senses and understanding were free of clouding fumes and animal spirits, Amerindians enjoyed lifestyles, including natural diets and monastic sexual habits, that cleared the brains of any obtrusive, clouding internal vapors.[71] Clearly, in Las Casas's version of things, Amerindians were among the most intelligent people on earth.

But colonists in Spanish America were not interested in arguments in defense of the Americas like those advanced by Las Casas. After all, these groups had long profited from representing the Amerindians as indolent phlegmatics who needed to be disciplined to get them to work. From the outset of the conquest, Amerindian forced labor became a vital institution of the Spanish empire.[72] The subsistence economy of local indigenous communities created shallow colonial labor markets and therefore the need for coercive mechanisms to supply ranches, textile factories (*obrajes*), mines, plantations, haciendas, and Spanish households with indigenous labor. Thus the ideology of the lazy native helped justify the institutionalization of labor coercion.[73] The construct of the phlegmatic Amerindian was a central ideological figure that allowed Spaniards to mobilize thousand of natives through the infamous forced labor system known as the *mita*. By the mid sixteenth century, Spanish colonialist knowledge had articulated a view of the Amerindian in which the Hippocratic writings and Galen were mustered to justify forced labor.

About 1570, Lope de Atienza (b. 1537) argued, for example, in a treatise on the natives of Peru addressed to Juan de Ovando (d. 1575), president of the Council of the Indies, that the Amerindians were phlegmatic and lacked beards owing to the excessive humidity of the New World. They were therefore unable to do manly things requiring "daring, constancy, and wisdom," namely, "graver things, things of better quality and standing than [those done by] women."[74] Their humid and phlegmatic constitutions made them not only effete but also idle, unwilling to work, and given to bouts of depression, curable only through harsh regimes of labor discipline.[75] These views were echoed down the centuries. By the early decades of the seventeenth century, Creoles wholeheartedly embraced the paradigm of the phlegmatic, effeminate Indian. The prestigious Spanish jurist Juan Solórzano Pereira (1575–1655), whose long stint in the Indies made him a staunch defender of Creole interests at the court in Madrid, made the paradigm of the phlegmatic Amerindian who should be forced to work widely available in the 1620s in the pages of his monumental *Política indiana*.

Solórzano argued that "the customs of each region are no less different than the air that bathes them," and that "the good legislator should always tailor his precepts to the regions and to the disposition and capacities of the people to whom they are addressed." Solórzano drew two different conclusions from this premise. "It would be a mistake to seek to instruct all [Amerindians] in the same way, for even during the time of their paganism they had as many petty kings and caciques as their constitutions and temperaments were varied," he asserted, on the one hand.[76] A single law for all Amerindians was not appropriate, because the climates of the Indies were varied. On the other hand, directly contradicting his first conclusion, and more important for the purposes of this chapter, he proposed that, since all Amerindians were phlegmatic, the Crown should follow the example of the

Aztec and Inca rulers, who had forced their subjects do things as absurd as to collect and fill bags with fleas to combat their innate idleness and keep them busy.[77] On such grounds, Solórzano supported institutionalized coercion and the *mita*.[78] A Spanish Jesuit long resident in the Indies, Bernabé Cobo (1580–1657) captured the biological discourse underwriting the ideology of the lazy Amerindian well in the mid seventeenth century: "All of them [Indians] are phlegmatic by nature, and since natural phlegm makes the substance of one's limbs soft and moist, their flesh is very soft and delicate. As a consequence, they tire easily and are incapable of working as hard as the Europeans. In Spain, a single man does more work in his fields than four Indians will do here."[79] But for all its positive ideological uses, the view that Amerindians were indolent phlegmatics also carried the assumption, implicit in the Hippocratic and Galenic theory of humors and constitutions, that negative bodily temperaments were caused by negative climatic and astral influences affecting everybody, not only the Amerindians.

Inventing the Racialized Body

Creoles and long-term European residents in the Indies were left facing an extraordinary paradox: how to maintain that America was under benign, soothing cosmic influences without giving up their construct of the Amerindians as phlegmatic miscreants. The works of León Pinelo and Salinas de Córdoba that I reviewed at the beginning of this chapter show the solution adopted by colonial intellectuals, that is, to postulate that Amerindians and Europeans had different sorts of bodies, which would make any radical transformation of the latter due to climatic or astral influences unlikely, if not impossible. This discourse permeated the works of the most important representatives of Creole patriotism in the seventeenth century.

How the discourse of two different types of bodies worked in practice can be seen in the writings of Antonio de la Calancha, who, wearing the hat of patriot astrologer rather than Augustinian friar, maintained that the skies of the Indies were protected by crosslike constellations whose soothing influences kept demons away and "pacified" the waters of the South Seas. I choose Calancha to typify the views of Creole patriots on the racialized body because he was one of the few Creole scholars in the century who explicitly denied that Amerindians were the cursed descendants of Ham, born to serve their European masters.[80] Yet, for all his efforts to present the Amerindians as vassals, rather than servants, of the Spanish monarchy, Calancha was forced to grapple with the paradox raised by his view of America as paradisiacal and providentially designed. If the continent were as temperate and under astral influences as benign as Calancha claimed, then it followed that Europeans, coming from colder climates, should be less intelligent than Amerindians. But Calancha, who could not stomach those who insisted Amerindians were servants by nature, could not even contemplate the possibility that they were in fact more intelligent than their European masters.

Calancha was aware that in Aristotelian theory, an idyllic representation of the New World implied of necessity that the temperate Amerindians were more intelligent than the colder Europeans. However, he denied that this meant a threat to the entire structure of contemporary medical knowledge, even though Aristotle's view that peoples from colder climates were less intelligent than those who hailed from warmer lands had been influential for a long time.[81] The heads of those who hailed from colder climates, according to Calancha's version of Aristotle, were deemed to be full of animal spirits, which, clouding the transfer, manipulation, and recovery of mental images in the five internal senses, hampered the correct operation of the understanding. Inhabitants of temperate climates, on the other hand, had less vital heat inside their bodies and therefore their heads were thought not to have clouding vapors limiting perception; they were expected to be more intelligent. Solely on the logic of this argument, Amerindians, the original inhabitants of the paradisiacal New World, should have been more intelligent that their colder European masters. Since Calancha was unlikely to be drawn to this possibility, two other alternatives were left. The first of these was to debunk the ancients and sign on to the effort of the moderns to invent and develop new understandings of the operations of the body. But joining Descartes's call to dismiss authority and create new theories in solipsistic bouts of philosophical meditations was something Calancha was not suited to. Rather, Calancha chose the more palatable second alternative of embracing Aristotle while at the same time negating the idea that climate and astral influences could induce profound transformations. The result of this operation to salvage Aristotle's paradigm, however, was a major break with ancient views of environmental determinism.

Although Calancha did not offer any details as to the composition of Amerindian, black, and white bodies, he insisted that within each group, Aristotle's dictum that peoples of warmer climates were more intelligent than those of colder ones held true. So according to Calancha, coastal Amerindians were smarter than Andeans, because the latter lived in colder climates; blacks born in Lima were more intelligent than their African parents, because Lima was more temperate than Africa; and finally Creoles were sharper than peninsulars, because Europe was colder than Spanish America. However, profound, innate racial differences distinguished the bodies of Amerindians, Creoles, and blacks, which rendered the transformation of one group into another impossible. The "materials" that made up the bodies of Amerindians, African blacks, and European white colonists were so different, Calancha argued, that different results should be expected from the operation of similar climatic and astral influences, in much the same fashion that the sun caused wax to melt, but clay to harden.[82] On this, Calancha cited the authority of Enrico Martínez (Heinrich Martin) (d. 1632), an émigré German polymath physician who in the late sixteenth and early seventeenth centuries participated in heated medical debates in Mexico on the nature and composition of Amerindian and Creole bodies. In order to understand Calancha's reference to Martínez, we need to turn back to late sixteenth-century New Spain, where the racialist paradigm was first hammered out.

Juan de Cárdenas wrote *Problemas y secretos maravillosos de las Indias*, arguably the first modern treatise on racial physiology, and a book whose originality and import have been overlooked, in Mexico City in 1591. Cárdenas presented his treatise as a survey of the secrets and marvels of the Indies, such as why there were frequent earthquakes, honey was sour, animals appeared not to suffer from rabies, and crawling creatures were less poisonous in the New World (all due to excessive humidity, in his view).[83] Cárdenas was also interested in human marvels, so he set out to explain why Creoles aged more quickly than their peninsular parents (the humidity of the Indies made them compensate and lose vital heat, and their bodies therefore desiccated sooner);[84] grayed faster (humidity dampened the body's vital heat, which in turn caused it to produce more phlegm than bile or blood, and since hair was made up of bodily excretions and phlegm was white, Creoles went gray faster);[85] suffered from more stomach problems (humidity diminished the natural heat of the stomach and prevented food from being cooked and digested well);[86] tended to be healthy when old but plagued by diseases when young (local humidity compensated for the lack of vital heat and dryness of the elderly but compounded the problems in youth);[87] and suffered inordinately from menstrual cramps (humidity created viscous substances that clogged the veins of a woman's uterus, which made the release of surplus blood painful).[88]

Notwithstanding all this, Cárdenas went to great lengths to argue that these humid-phlegmatic defects should not be confused with the Amerindian constitution. The problems of white people in the Indies were caused by an "accidental" phlegmatic constitution, an environmentally induced superficial quality, added to a naturally sanguine temperament. The phlegmatic constitution of the Amerindian was, on the other hand, "natural."[89] Cárdenas's use of the terms "natural" and "accidental" was not casual;[90] it derived from a convention he and his readers took for granted, namely, Aristotle's metaphysics. In Aristotelian parlance, the word "natural" refers to the predictable behavior of objects and organisms, whereas the word "accidental" refers to the passing properties of "forms." Substantial "forms" according to Aristotle give matter its changeless attributes, whereas "accidental" ones explain change and refer to passing characteristics of objects.[91] Clearly, the reference in Cárdenas to "natural" constitutions conveyed the notion of innateness. According to Cárdenas, the environment could not change "natural" racial traits and effect profound bodily transformations that would lead Creoles to mutate into Amerindians; astral influences and climate operated within limits set by innate bodily differences.

Cárdenas was hard pressed to maintain his essentializing, reified bodily categories, and contrived unpersuasive theories to do so. To explain why Amerindians never went gray whereas Creoles did, he argued, for example, that Amerindians, although phlegmatic in "nature," lacked excess "accidental" phlegm, the origin of gray hair.[92] The question of why Amerindians were beardless and never went bald, whereas Creoles lost hair as they aged but had abundant beards, prompted him to ask why they should differ in this way if both groups lived in the same environment, ate the same food, and drank the same water. This question was only relevant, however, if one assumed, in Cárdenas's words, that "the composition and organization of our [body] and theirs is one and the same."[93] This clearly was not the case. The differences between Amerindian and Creole bodies were profound, Cárdenas thought, and he offered strikingly different views of them. Creoles went bald as they lost vital heat (because they aged sooner), the skin on their skulls became tighter and dryer and obstructed the release of hair, a bodily extrusion. The Amerindian was phlegmatic by nature and bound always to stay humid; he therefore never lost hair. Spanish Americans, due to their sanguine and choleric constitutions, were tight and dry, with a patina of "accidental" phlegm added to them by the humid climate of the Indies. The Amerindians, on the other hand, were moist and phlegmatic by nature.[94] The humid character of the Amerindians made them effete and therefore caused them to be beardless.[95] The only advantage Amerindians had over their European masters was that, due to their lack of "accidental" phlegm, they did not suffer from joint aches (*reumas*), urinary tract infections (*males de orina*), gallstones (*males*

de ijada), and stomach problems.[96] Cárdenas did not discuss the possibility that the "accidentally" humid-phlegmatic yet "naturally" dry-choleric
Creoles were becoming Amerindians. Creoles, Cárdenas maintained, had
been born to peninsular parents whose temperament was choleric, but they
were slightly transformed in the Indies to acquire an even better sanguine
constitution, which gave them better mental qualities than their parents
born in colder climates. Creole brains were freer of clouding animal spirits,
allowing them to manipulate and retain mental images in their internal
senses better than their European parents.[97]

Cárdenas's theories were to reverberate in Mexico in the years to come when
two émigré physicians, one German, one Spanish, engaged in heated debates
over how to determine the stars and constellations that ruled New Spain and
the constitutions of its peoples. In 1606, hired to oversee the works to drain
the capital, Enrico Martínez contended that Mexico was under the influence
of Capricorn, and that plagues were likely when a conjunction of Mars and
Saturn occurred under this sign of the zodiac.[98] Since 1519, Mexico had witnessed three such planetary conjunctions and hence had suffered three devastating epidemics, which had decimated the indigenous population, Martínez
noted, but these plagues had spared most Spanish colonists. To explain this,
Martínez drew on Cárdenas, arguing that Creoles and Amerindians had different constitutions and therefore different physiological reactions to similar
climatic and astral influences. Saturn and Mars were warm and dry, conflicting with the cold, humid constitutions of the natives, and therefore triggered
a surge of bile in them, which led to the Amerindians being wiped out.[99]
Martínez introduced the same distinctions between natural and accidental
constitutions used by Cárdenas a few years earlier. According to Martínez, the
Amerindians were "naturally" phlegmatic (and thus prone to be debilitated
by the conjunction of Saturn and Mars under Capricorn) and "accidentally"
sanguine. Since, according to Martínez, Mexico was ruled by Venus (which
stimulated the production of phlegm) and the sun (which stimulated the production of blood), the natives were doubly phlegmatic due to the added effect
of Venus on their essential nature, yet they were also slightly sanguine due to
the influence of the sun.[100] Creoles, however, were immune to the debilitating
phlegmatic influences of Venus, because they were essentially choleric (cho-
ler, or yellow bile, was able to counteract phlegm); they responded only to the
beneficial sanguine influence of the sun. Like Cárdenas, Martínez asserted
that the Creoles' choleric inheritance had turned sanguine in Mexico under
the benign planetary influence of the sun. Creoles were more intelligent than
their Spanish parents, although weaker.[101]

Martínez maintained that positing such radical racial differences confuted critics of America who argued that "if the qualities of these lands are suited to create good minds, those native to the land should enjoy very good ones . . . so much so that they [the Amerindians], and also blacks, should be as intelligent as the Spaniards, for all of them share [the qualities of the environment] equally. Experience, however, proves the opposite because these peoples [Amerindians and blacks] have [mental] abilities far inferior to [those] of the Spaniards. One therefore has to conclude that this land has none of the properties alleged by its supporters."[102] Martínez identified the crucial paradox with which Creole patriots had to grapple for the rest of the century. How could a superior land, America, be home to Amerindians and blacks, believed to be far inferior to their European masters?

Martínez responded to the arguments of the imagined critic by insisting that the properties of place could not be judged by the quality of its peoples. "Universal causes," Martínez argued, "are modified according to the quality of the matter [upon which the climate acts], influencing diverse subjects differently; fire burns both dry and green wood, but the former far faster than the latter. The constitution of blacks is very different from that of Spaniards, [and] so too is that of Indians. That is the reason why similar general influences in this kingdom do not produce the same effect in all [populations], but according to the temperament, disposition of the brain, and body organs [of each]. From this proceeds the diversity of the genius of the said [three] nations."[103] So, according to Martínez, members of each of the three "nations," Amerindians, whites, and blacks, might be expected to change according to the climate in which they lived: peninsulars who emigrated to the Indies and their offspring enjoyed better constitutions because they had moved from colder to more temperate climates; blacks born in America were vastly superior to those of Africa; and the Amerindians of New Spain were more civilized than those of the Caribbean and Florida.[104]

A few years later, the views of Martínez were paradoxically reinforced by Diego Cisneros, an émigré Spanish physician. Paradoxically, because in 1618 Cisneros published a book, *Sitio, naturaleza y propiedades de la ciudad de México*, in part to take issue with Martínez's astrological characterizations of the lands and peoples of Mexico. Cisneros argued that Mexico was not ruled by Capricorn, Venus, or the sun, as Martínez had maintained.[105] To determine the ascendant stars of Mexico, Cisneros averred, two elements were needed, the date of foundation of the capital and its location (both longitude and latitude), and Martínez had failed to provide either one correctly. According to Cisneros, Martínez had wrongly located Mexico City on the map. Moreover, Martínez, who assumed that the ascendant stars of a given location were those that happened to be over the place at the time of creation, had miscalculated the date of creation. Cisneros maintained that it was impossible to determine the date of creation, and that therefore it was impossible to find Mexico's ascendant planets.[106] Yet Cisneros, who sought to emulate Hippocrates in America by writing a treatise in which health and disease were explained as the outcomes of temperaments and bodily humors governed by the quality of food, water, locale, and wind patterns, did not disagree with Martínez regarding the radical innate bodily differences between Creoles and Amerindians. Cisneros rejected Cárdenas's and Martínez's contrived theories postulating "accidental" and "essential" constitutions and ridiculed Martínez for suggesting that the Amerindians could be both phlegmatic and sanguine. But Cisneros did not reject the essentializing and reifying distinctions between Amerindian and Creole bodies that Cárdenas and Martínez had introduced. Like them, Cisneros drew a clear racial line and argued that Amerindians were melancholy (i.e., cold and dry, as opposed to the cold, wet temperament of the phlegmatic constitution) and Creoles sanguine.[107] Also like Cárdenas and Martínez, Cisneros maintained that climate worked within the limits established by set racial differences and concluded that the climate of the America had transformed the choleric Spanish inheritance of the colonists into an exceptionally good, sanguine constitution.[108]

The hypothesis early in the seventeenth century by three long-term European residents of the Indies that Amerindians and whites had different types of bodies (although blacks were often mentioned, there was no attempt to clarify their essential and accidental constitutions) was part of a broader cultural response to negative European views of the climate and heavenly constellations of Spanish America. Europeans, Amerindians, and blacks were perceived as essentially different; their "temperaments, the disposition of their brains and body organs" were thought to be made of different "matter" and thus expected to react to the same environment differently. Martínez's theories were repeated by Creole patriots throughout the seventeenth century. Some ninety years after the publication of Martínez's book, the Franciscan friar Augustín de Vetancurt (1620–1700) could still quote him verbatim. In his *Teatro mexicano*, a work published in Mexico in 1697–98, Vetancurt insisted that Mexico was a microcosm of the world, providentially designed and chosen to nurse the globe with silver and gold rather than milk. Seeking to counter the skeptical claims of those who maintained that America could not possibly be such a paradise, given that inferior Amerindians and blacks had been born there, Vetancurt turned to Martínez and argued that "universal causes are modified according to the quality of the matter [upon which the climate acts]; fire burns both dry and green wood, but the former far faster than the latter. The constitution of blacks is very different from that of Spaniards, [and] so too is that of Indians. That is the reason why [climate and heavenly constellations] in this land do not produce the same effect in all populations; it varies according to the temperament, disposition of the brain, and body organs of each."[109]

The hypothesis that there were two (or more) distinct types of human bodies might also, paradoxically, have been a result of the racial complexity that came to characterize the Spanish American colonies in the early seventeenth century. What is puzzlingly absent from the racialized view of the body that I have discussed here is any reference to miscegenation or to the bodies of *castas*. As R. Douglas Cope has shown, it was precisely when the authors I have discussed here were writing that mestizos and mulattoes were becoming the majority of the colonial urban population, far outnumbering both Amerindians and Spaniards. This situation led the elite to introduce the infamous *casta* system of racial distinctions in the hope of disciplining and controlling a class of urban plebeians perceived as undifferentiated and threatening. But for all the fine racial taxonomies seeking to "divide and rule" that were introduced over time (forty by some counts), the *casta* system never put down firm roots, and only two *casta* groups had a lasting, meaningful existence, mestizos and mulattoes. This tendency to simplification, Cope has argued, was the result of resistance on the part of the poor urban majorities to racial labeling. But the simplifications might well have originated in the inability of Spaniards and Amerindians—precisely those the system was intended to benefit—to think beyond sharp racial typologies.

The reaction of Spaniards and Amerindians alike to the increasing racial complexity of the colonies was to fall back on polarized views of society. Just at the time when Martínez was arguing that there were three separate types of bodies, Don Domingo de San Antón Muñón Chimalpahin Quauhtlehuanitzin (1579–1660), a Nahua annalist and a self-styled member of the indigenous nobility of Chalco, wrote in Mexico City that "at the foundation and beginning of the world, we had only one father, Adam, and one mother, Eve, from whom we descended, although our bodies are divided into three kinds."[110] Miscegenation paradoxically prompted both the Spanish and Amerindian elites to simplify and polarize racial divisions within the colonial polity. This polarized view of bodies and society, as Nils Jacobsen has shown, should perhaps be considered one of the most significant colonial legacies of modern Latin America, a legacy that, at least in Peru, has been largely responsible for a thwarted and incomplete transition to a capitalist market economy.

Conclusions

Some scholars have argued that the science of race originated in western Europe in the mid eighteenth century among natural historians seeking to devise ever more sweeping taxonomies to encompass the bewildering variety of living things that confronted them in the aftermath of the age of exploration. Others hold that it arose in the nineteenth century as a result of the major political transformations of the age of revolutions. As the old political orders based on social estates, hereditary privileges, and religion came tumbling down, and new social formations founded on the principles of citizenship, natural rights, and secular political authority emerged, white European males located the ideological justification for preventing women, slaves, and non-Europeans in general from sharing their own newly acquired political rights in discourses of race and sex.[111]

Historians of colonial Latin America in particular are entitled to be skeptical of these accounts of the genesis of the science of the racialized body. Learned sixteenth-century Spaniards brought to the New World ideas and theories that at first glance could be interpreted as forms of the science of race. Spanish humanists and neoscholastic theologians engaged in long-drawn-out debates (the focus of sustained scholarly attention today) in which the natives of the New World were portrayed as natural slaves, peoples whose brutal ways and elementary misunderstanding of even the most basic natural laws showed that they lacked reason and political judgment, and who therefore had no right to hold property and rule themselves. This paradigm, which helped justify early Spanish colonial expansion, was soon superseded by a view of the Amerindian as possessor of a human soul and thus a potential member of a Christian commonwealth. In this new version, however, the Amerindian was represented as a psychologically arrested child, whose innate natural rights had to be administered by proxy. Cast in the idioms of medieval and Renaissance natural philosophy, these views created the essentializing and homogenizing category of the "Indian" in European consciousness.[112]

Yet for all their similarities to the modern concept of race, the views of the natives as natural slaves or as childish barbarians were not predicated on the study of the Amerindian body. Spanish colonialist knowledge created categories that classified peoples according to whether they had cities, knew how to build arches, enjoyed hierarchical social arrangements, created systems of writing, practiced the right religion and rituals, knew appropriate dress codes, and understood dietary and sexual "natural laws."[113] The secular concern of early modern Spanish Christians with keeping their blood pure and uncontaminated by threatening neighboring religions (Jewish, Muslim, and, later, Protestant and native American) may at times give modern observers the impression that in the Spain of that era, cultural differences were in fact explained by claiming that there were essential biological differences between nations. The discourse behind the concept of "purity of blood" maintained that idolatry was a behavior passed down through the generations by corrupted blood.[114] Spanish writers thus often explained the refusal of American natives to abandon their ancestral religions as a physical trait of the latter, a "bad seed that has grown deep roots and has turned itself into blood and flesh [in the Amerindians] . . . a vice that comes in the blood and is suckled as milk [from their mothers' breasts] . . . [because] the customs of parents and ancestors are converted into nature and transmitted through inheritance to their children."[115]

A closer look at these views shows that nurture, not nature, was the dominant explanation offered for religious deviancy. Culture, the customs of parents, literally transformed the body, modifying the blood and flesh of Amerindians. As Anthony Pagden has argued, it was the assumption that culture could become so ingrained as to become a second human nature that led sixteenth-century Spanish scholars to argue that it would take a great many generations of hard work by missionaries to transform the psychologically arrested, childish Amerindians into full-fledged, Christian European adults. Nonetheless, Amerindian conversion and Europeanization was deemed possible.[116]

By the early seventeenth century, however, this faith in the transforming (and redeeming) power of evangelization and acculturation gave way to marked skepticism in Creole learned circles. As Sabine MacCormack has suggested, seventeenth-century Creole authors, witnessing the indigenous refusal to give up ancestral religious practices lightly, embraced racialist views to explain religious deviance among Andeans. In previous centuries, Christian scholars had maintained that the devil was largely responsible for idolatrous behavior, because Satan misled worshippers through the manipulation of people's mental faculties, making them see false phantasms and religious visions (the devil also altered the actual physical environment). Responsibility for religious visions thus lay outside the bodies and minds of idolatrous worshippers. This paradigm was used extensively in early colonial Spanish America, but by the seventeenth century, the Creole clergy in the viceroyalty of Peru began to blame Amerindian idolatry on the flawed operation of the Amerindian body, rather than on the machinations of the devil. Indigenous religious deviance thus became a psychological problem, largely attributable to the physical malfunctioning of the internal senses of Amerindians, the failure of their brains to grasp the logical, scientific structure of the universe.[117]

The works of the two seventeenth-century scholars Salinas de Córdoba and León Pinelo, with which I began this chapter, are representative of the efforts of the learned in Spanish America to invent two different bodies, one for Amerindians and the other for white European colonists (and, in lesser degree, a third one for African blacks). As they sought to defend themselves against a rising European intellectual current that portrayed the climate and heavenly constellations of Spanish America as causing mental and physical degeneration, Salinas and Pinelo, along with a cadre of other Creole and émigré European scholars, hammered out forms of patriotic astrology, and, more important, a science of the racialized body that long predated that invented in the late eighteenth and early nineteenth centuries in Europe. The "modern" science of race was thus first articulated in a "peripheral" colonial setting and worked out using distinctively nonmodern idioms. Yet, paradoxically, this very modern creation, it seems, never had much of an impact outside Spanish America. Although the colonies functioned, in Ann Laura Stoler's terms, "as laboratories of modernity," the fact that modernity defined itself by dismissing colonial intellectuals and by denying the status of science to astrology, Aristotelian metaphysics, and Galenic physiology doomed these first modern views of the racialized body to become invisible in European consciousness.

CHAPTER 5

Eighteenth-Century Spanish Political Economy
Epistemology and Decline

As we saw in Chapter 2, references to the backwardness and ignorance of Spain extracted from the pages of scores of eighteenth-century travelers and authors would most likely fill a thick volume. The European Enlightenment had no patience with Spain, for it stood for all the things the literati most hated. These criticisms, many of which were broadcast not only by overzealous ideologues but also by important scholars such as Montesquieu, left a lasting mark on the soul of the Spanish Enlightenment. The state and local intellectuals spent considerable amounts of energy seeking to deny the charges leveled by foreigners, particularly the French. There is no better example of how anxiously Spaniards sought to refute these negative images than the cultural and scientific policies undertaken by the Crown. To counter the views of the French Minim Louis Feuillée, who in the wake of expeditions to Peru and the Caribbean early in the century maintained that the New World remained largely unknown, owing to Spain's lack of scientific curiosity, and to rebuff Carl Linnaeus (1707–78), who complained that Spain had done nothing to improve the knowledge of plants, the Crown poured millions into building botanical gardens and outfitting countless natural history expeditions to the colonies. Although this project did not yield concrete economic returns, it was so massive that by the end of the century, Alexander von Humboldt could not contain his surprise. Enlightened botanists such as Antonio José Cavanilles used the generous policies of the Crown to name new plant species after prominent local men in order to remind the European public of the accomplishments of the Spanish literati.[1]

Neither criticism of Spain's alleged backwardness nor Spanish patriotic responses to foreign scorn were new in the eighteenth century. There was, however, one novel development. Whereas censure had in the past been focused on the alleged idolatrous and superstitious nature of Spanish Catholicism and on the brutality and greed exhibited by conquistadors and troops in America and Holland, by the mid seventeenth century, European intellectuals found themselves grappling with the reality of Spanish "decline."[2] Travelers and observers began to hammer out a cautionary morality tale of a nation in precipitous freefall, a country that had gone from riches to rags, squandering opportunities at every turn. This master narrative maintained that the spread of aristocratic values among the laboring population, the irrational rejection of the new sciences and technologies developed in the rest of Europe during the Scientific Revolution, and the building of a fiscally irresponsible, overstretched empire had condemned Spain to a chronic state of depopulation and vagrancy.[3]

Yet in all these neat narratives of decline, causes and effects were difficult to sort out. Was idleness, for example, an innate Spanish trait resulting from the North African climate of Castile, as many alleged? Or was it rather a product of the spread of aristocratic values among populations long locked in a crusading campaign against Islam? Was unemployment actually caused by the massive arrival of cheaper foreign goods and textiles, which since the late sixteenth century had driven much Spanish manufacturing out of business? By the same token, was the disproportionate growth of the clergy caused by the search for alternative forms of employment? Or was the swollen clerical state the reason for the poorly performing economy in the first place, inasmuch as the clergy had promoted a culture of profligate ritual excess, ignorance, and lack of entrepreneurship?

Early modern debates over the cause of national decline began first in Spain. In the early seventeenth century, the country witnessed an explosion of so-called *arbitrista* literature (proposals as to how to solve fiscal or sociopolitical problems) that sought to explain why Spain had so rapidly lost ground to other European powers. The remedies for decline formulated by some *arbitristas*, however, ultimately proved detrimental. Like other European intellectuals bent on explaining the rise and fall of empire, many *arbitristas* blamed courtly emasculation for the current plight of the nation. Not surprisingly, they pressed for a return to Christian piety and advised renewed military campaigns to bring out the warrior in Spain's effete courtiers. To pay for these wars and prevent excessive luxury, they proposed sumptuary taxes. Such measures, however, only worsened the fiscal deficit and further slowed down the economy.[4] No wonder, then, that the *arbitristas* became figures of ridicule and scorn in seventeenth-century Spanish theater.[5] Eighteenth-century literature cautioned repeatedly against the example of the *arbitristas*, whose misguided theories had further compounded the crisis. Thus, it was the patriotic responsibility of the intellectual to identify the cause of decline accurately. But for all the realization that many *arbitris-*

tas had done more harm than good, Spanish intellectuals remained wedded to traditional discourses. It was difficult for new paradigms to spread.

Since the Renaissance, traditional discourses had held that decline was caused by the loss of civic virtues. Theories of the decline of Rome were emblematic of this approach. Decadent, self-indulgent elites, it was argued, had left the defense of the Roman empire in the hands of professionals. As commerce and territory grew, and as citizens became merchants or, worse, idle, effete courtiers, the Romans had turned over the affairs of the state to mercenaries and emperors. Soon the Romans had caved in, unable to take power away from emperors and cravenly surrendering to the barbarian onslaught.[6]

This discourse of decline through emasculation came in for criticism, however, as the eighteenth-century commercial revolution unfolded. As absolute monarchies consolidated, and as mercantilism and new cultures of sociability and consumption came of age, a new discourse of "commercial humanism" replaced the theories of decline resulting from trade and luxury. The views of Bernard Mandeville and Adam Smith typified this transformation. In *The Fable of the Bees* (1714), Mandeville argued that private "vices" were, in fact, public virtues; luxury and the pursuit of wealth benefited society as whole. Adam Smith's doctrine of the "invisible hand" also held that self-interest became civically virtuous because of the law of unintended consequences, a form of indirect providence. Theories of checks and balances limiting power helped scholars to account for decline. Lack of parliaments, not excessive commerce, caused nations to decline.[7] In Spain, however, patriotic epistemology and opposition to new forms of sociability and consumption slowed down the spread of the discourse of commercial humanism.

In the following pages, I describe how the main preoccupation of eighteenth-century Spanish political economy was the identification of the causes of national decline. The discipline developed in full awareness that European discourse scathingly critical of things Spanish had already framed the debate. Spanish intellectuals dismissed foreign observations, claiming that they alone were capable of being objective on the subject. Patriotism, however, narrowed the choices available to political economists to theories that explained decline in terms of luxury and emasculation. Many intellectuals found the virile and humanist culture of the Spanish Renaissance (*siglo de oro*) an alternative to the new French forms of courtly and public sociability introduced by the Bourbons. But traditional theories regarding the negative effects of luxury on national prosperity had to confront the

record of some *arbitristas*. As eighteenth-century Spanish political economists realized that traditional paradigms could further damage the economy, they engaged in endless debates about the causes of economic decline,
the identification of which became a patriotic crusade. Two clear positions
emerged from this debate. The first was that the causes of decline were
already well known, and its proponents advised immediate reform along
the lines suggested by the much-maligned foreign observers. The second
advocated historical research, contending that shallow theorizing had led
Spain astray, causing more economic damage than stability. Advocates of
this second position maintained that instead of speculative theorizing, careful archival research into the historical processes of decline was necessary.
Political economy in Spain developed along these two mutually exclusive
tracks.

Patriotic Epistemology and the Critique of New Forms of Sociability

It is plausible to argue that a fundamental feature of the Spanish Enlightenment was the tension that obtained between rejection of scornful foreign
visions of Spain and the need to accept the reality of decadence while identifying its causes.[8] The critique of foreign views was a central concern of the
age. Foreign commentators were dismissed because they were thought to
pass judgment on a nation they barely knew. Spanish intellectuals characterized foreign observers as shallow and ignorant, lacking the linguistic and cultural tools needed to make sense of the country they purported to interpret.
The Valencian Juan Bautista Muñoz typifies this attitude. When Cesareo
Giuseppe Pozzi, a member of the entourage of the papal nuncio in Madrid
and former professor of mathematics at the Sapienza in Rome, published
his *Saggio de educazione claustrale per li giovani* (*Essay on the Education
of Cloistered Youth*), which ridiculed Spanish clerical training, in 1778,
Muñoz immediately wrote a scathing critique, castigating the philosophical and theological views underpinning the Italian's educational proposals.
More important, Muñoz set out to prove that the erudite Benedictine was
in fact a fraud, who had lifted most of his entries on pedagogy from fashionable French texts, assuming that Spanish audiences were too benighted to
notice. A foreigner like Pozzi felt entitled upon arrival to pontificate on matters he barely understood simply because he held Spaniards in contempt.
Pozzi, Muñoz argued, typified the European intellectual who, "without any
knowledge of Spanish learned traditions, having most likely read a few of our

books we ourselves dismiss, pigeonhole us as ignorant, as defilers of good taste."[9] Muñoz was not alone. In 1782, reviewing Henry Swinburne's *Travels Through Spain, in the Years 1775 and 1776* (1779), José Nicolás de Azara, Spain's powerful ambassador in Rome and a leading patron of the Spanish Enlightenment, also broadly criticized foreign travelers. "Swinburne," Azara mockingly remarked, "should be praised for his sharp observational powers, for after having spent two, three days [in Spain] he already knew that all roads were terrible, hostels even worse, and the whole country a hellhole where stupidity runs rampant."[10]

The condemnation of ignorant foreign observers reached a climax in the wake of the publication of Nicolas Masson de Morvilliers's article "Espagne" in the *Encyclopédie méthodique* (1783). An author more given to rhetorical flourishes than to serious scholarship, Morvilliers caused an uproar after he argued that Spain had contributed nothing to the store of knowledge in the past millennium. In 1786, in his *Oración apologética por la España y su mérito literario* (Apologetic Peroration on Spain and Its Literary Merit), Juan Pablo Forner denounced these views as typical of the French, whose liking for grand philosophical systems such as that of Descartes made them fond of outlandish sweeping statements. Morvilliers was representative of foreign observers "who after having simply perused our annals, having never read our books, clueless of the state of our higher education, ignorant of our language, prefer not to draw upon the proper sources of instruction but on what comes more easily, namely, fiction. Thus, at the expense of our sad peninsula, they spin off fantastic novels and fables as absurd as the stories retailed by our early writers of books of chivalry."[11]

For all their rejection of foreign scorn, Muñoz, Azara, and Forner were perfectly aware that Spain had lost ground to rival powers. These authors spent their lives identifying the origins of the decline and seeking solutions. Yet they also thought that Spaniards, or expatriates who had made Spain their home, were better equipped than foreigners to tackle this difficult intellectual problem.[12] This patriotic epistemological discourse lies at the core of the Spanish Enlightenment and gives the period its peculiar texture. This discourse surfaces repeatedly throughout the century and helps explain the work of José Cadalso, one of the leading intellectuals of the age.

Cadalso's *Cartas marruecas* (Moroccan Letters) (1793) is paradoxically a work that privileges the views of a foreign observer. Employing the device of fictional letters penned by a Muslim traveler, a conceit introduced by Montesquieu in *Lettres persanes*, Cadalso uses the Moroccan Gazel to discuss the dilemmas and problems of contemporary Spanish society.

Inasmuch as he made Gazel central to his narrative, it would appear that, unlike Azara, Muñoz, and Forner, Cadalso was not interested in discrediting foreign reporters. But Cadalso held fast to patriotic views of epistemology. Gazel is credible precisely because he is always ready to listen and to learn from Spaniards. Throughout the text, Gazel comes across as eager to be taught and be educated by Nuño Núñez, a deeply learned Spaniard well acquainted with the sources of national history. Moreover, Gazel speaks fluent Spanish and embraces Spanish sartorial fashions. Ultimately, it is Núñez's voice that Gazel parrots. Núñez (representing Cadalso in the text) uses Gazel to vent his criticisms of Spain.[13]

Cadalso was a fierce proponent of patriotic epistemology. In fact, his *Cartas* were first conceived to address the criticisms leveled against Spain by Montesquieu. In his *Lettres persanes*, Montesquieu had attributed Spain's stagnation to the exaggerated Spanish culture of honor that seemed to value idleness over labor, a stance he portrayed as inflated and ridiculous. Cadalso was at a loss to explain why a man as learned as Montesquieu had dared to offer such sweeping and shallow generalizations of a nation he barely knew. Cadalso could not understand why Montesquieu had chosen to speak "without any knowledge of our history, religion, laws, mores, and nature." When Cadalso first read Montesquieu's letter no. 78, he became so enraged that he spent several years (1768–71) writing a *Defensa de la nación española* (Defense of the Spanish Nation).[14]

This defense not only called into question the credibility of foreign observers but also sought to offer an alternative account of the origins of national decline. Cadalso dismissed Montesquieu's views outright, arguing that Spain's problems did not arise from the culture of Castilian honor but from an aging, stagnant economy, shallow job market, and lack of avenues of social mobility. Cadalso insisted that this stagnation had started with the reign of Philip II, that "prejudicial king" (*un rey perjudicial*) whose rule had created the conditions that made the vagrancy and culture of idleness in Spain possible. According to Cadalso, Philip II's stubborn and fiscally ruinous defense of an overstretched empire was to blame for the crisis of productivity that engulfed Spain in the seventeenth century.

Cadalso's ferocious epistemological patriotism forced him to fall back on traditional *arbitrista* paradigms. In opposition to Montesquieu, he defended the culture of Spanish honor and gravitas. Cadalso's political economy assumed honor and manliness to be the engine of Spanish history: when these were scarce, the nation declined; when abundant, it prospered. Cadalso understood the Spanish past as cyclical, with periods of growth followed by decline. Phoenicians, Romans, and Visigoths, he noted, had at different times dominated Spain. These virile invaders had found a bountiful land and flourished. Soon, however, prosperity turned into decline and manliness into emasculation. New waves of virile invaders displaced the earlier ones, now turned weak and effete: the Phoenicians fell to the Romans, who gave in to the Visigoths, who in turn surrendered to the Moors. According to Cadalso, the Reconquista was part of this cyclical pattern of conquest, expansion, and decline. The effeminate Moors had been defeated by the more rugged Christian troops, who embarked on a new cycle of prosperity-cum-expansion, which had reached its apogee in the Renaissance. However, the reign of Philip II had brought prosperity to a halt and once again begun an age of "absolute decline in the sciences, the arts, the military, commerce, agriculture, and population growth."[15] Under the Habsburgs, traditional Spanish honor and virility became a mockery of themselves; the nation came to be led by buffoons and the mentally ill.

Like some *arbitristas*, Cadalso identified luxury resulting from commerce as a threat. However, unlike these *arbitristas*, Cadalso did not seek to enhance the nation's virility by taxing it to death or by promoting new military campaigns. His theory of decline rather led him to oppose the new forms of sociability sponsored by the Bourbons. Cadalso detested the Francophile character of Spain's new public sphere, which he found decadent and fake. In his efforts to castigate this culture, he contributed to the creation of a new literary figure: the *petimetre*, a shallow and effete character who aped French fashions and who was given to vacuous and affected conversations. In *Los eruditos a la violeta* (1770) and *Cartas marruecas*, Cadalso presented a ferocious critique of the new civilization of sociability and consumption that seemed to be eating away at the virile soul of the Spanish nation.[16]

Seeking to halt the spread of shallow new foreign forms of sociability, intellectuals denounced the moral corruption they seemed to be inducing. Clearly, Cadalso was not the first to make this connection. Diego de Torres Villaroel had already maintained that the new forms of sociability were manifestations of corruption and symptoms of decline. An author whose distinct and powerful voice has yet to be understood, Torres Villaroel turned for inspiration to the seventeenth-century writer Francisco Quevedo, who had masterfully deployed irony to defend Spain against foreign scorn. In *Visiones y visitas* (Visions and Errands) (1743), Torres Villaroel follows Quevedo on a fictional tour of Madrid, only to find the capital sunk in decadence. His pen portraits of dissolute and ignorant courtiers, gluttonous priests, swindling doctors, and writers and *petimetres* engaged in vacuous banter in the pursuit of fleeting fame, are all literary jewels. They also reveal a deep hostility to the new forms of public sociability embedded in the rising culture of *tertulias*, salons, academies, and cafes of mid-eighteenth-century Spain.[17]

This critique of the shallow new culture of literary sociability went hand in hand with a positive evaluation of the Renaissance as a golden age. Cadalso, again, exemplifies this trend. He turned to the Spanish Renaissance to help him deflect Montesquieu's criticism of the accomplishments of the Spanish mind. Montesquieu had argued that Spain's intellectual production did not go beyond books of chivalry and boring scholastic tomes. Cadalso replied to these charges by insisting that early modern Spain had been a nation intellectually more precocious than France, producing works of rhetoric, navigation, cartography, poetry, mathematics, and theology.[18] The revival or invention of the Spanish Renaissance as a patriotic strategy to counter foreign charges of Spanish ignorance and backwardness occupied the energies of scores of eighteenth-century intellectuals. Thus, in the words of a leading advocate of neoclassical reform in drama, Juan de Iriarte, a central objective of the Spanish Academy created by the Bourbons in 1712 consisted of "drafting apologetic defenses of the national language to counter the slanders of foreigners. Some rob us of our national pride, confusing us with Africans or Asians. Others cannot find among us more than one or two good authors and maintain that our science is limited to a couple of sonnets and a handful of syllogisms. [The Academy] needs to praise the great luminaries of our fatherland by reissuing their works and resurrecting their names."[19] These initiatives proved remarkably successful. Hundreds of early modern manuscripts and books by well- and little-known authors were either reissued or published for the first time in the eighteenth century.

For Cadalso and many others, the visions, styles, and sensibilities of the Spanish Renaissance stood for the "authentic," a means to oppose fashionable new intellectual currents that were foreign, effete, shallow, and fake. According to these intellectuals, addressing the problem of decline involved first and foremost the implementation of new cultural policies. As Antonio Mestre has repeatedly shown, the most representative figure of this type of approach was Gregorio Mayans y Siscar, who urged his countrymen to imitate exacting, elegant, and profoundly "authentic" scholarship of the Spanish "Golden Age." Many were inspired by this vision.[20] The very category of *siglo de oro*, a label now used to refer to a period in the literature of the seventeenth century, was first coined in the Spanish Enlightenment to describe the intellectual accomplishments of the sixteenth century.[21]

Chapter 5

Epistemology and Spanish Decline

In addition to concerns over the rise of new forms of sociability and consumption, discourses on decadence were framed by debates about whether the causes of Spain's decline were known and how fast to enact reforms. Many scholars, however, concluded that it was the very obsession with finding causes that accounted for Spain's decline in the first place. In the following pages I demonstrate how these concerns help make sense of key eighteenth-century Spanish texts on political economy.

The views of Manuel Antonio de Gándara on the causes of decline were typical of the debates of his age. In *Apuntes sobre el bien y el mal de España* (Notes on the Strength and Ills of Spain) (1759), Gándara showed that at least twenty-seven contradictory models could be cited to explain Spanish decline. So many models, however, obscured rather than clarified the real causes of decadence. Facing so many narratives of decline, Gándara argued, the authorities could not spring into action to enact much-needed reforms.[22] Gándara thought that political economy should investigate how all these seemingly unrelated models were in fact connected. Many *arbitristas* and early eighteenth-century writers such as Gerónimo de Uztariz, the author of *Theórica, y práctica de comercio y de marina* (1724, 1742), he complained, had correctly identified the symptoms yet had failed to discover the root cause of all Spanish ills, namely, the policy of allowing unchecked foreign imports, while actively thwarting local manufacturing. The inability of most *arbitristas* and previous political economists to pinpoint the cause of all Spain's symptoms had in itself become a problem for the economy. By seeking to treat the symptoms as opposed to the actual disease, these intellectuals had further injured the body politic.[23]

León de Arroyal begged to differ. In his *Cartas económico-políticas* (Letters on Political Economy) (1790s), Arroyal took issue with writers such as Gándara. According to Arroyal, the causes of Spain's decadence were multiple, interlocking, and yet to be fully identified (e.g., the cumbersome and archaic structure of the state; the many regulations choking both commerce and the treasury; the culture of litigation, undermining the implementation of the law; and the absence of checks and balances limiting both centripetal, authoritarian monarchical forces and centrifugal aristocratic ones). For Arroyal, the task at hand was the opposite of that urged by Gándara. Political economists should seek to discover the causes of Spain's decline through empirically and historically disciplined research. "Until the causes of Spanish decadence are discovered," he insisted, "all efforts to halt its devastating manifestations will prove in vain."[24]

Narratives of decline seem to have oscillated between the two alternatives sketched by Gándara and Arroyal. On the one hand, there were those who sought to create a deductive science in which all the symptoms of decline were accounted for by a single all-encompassing theory. On the other hand, there were the likes of Arroyal, who understood political economy as a historical science and feared that hasty theorizing might compound the crisis by introducing misguided reforms. There was, however, a third position, first sketched by Melchor Gaspar de Jovellanos, one of the leading figures of the Spanish Enlightenment.

Jovellanos has been the focus of much sustained scholarly attention, yet scholars have failed to notice that the treatise in which Jovellanos laid out his program of liberal reform, the *Informe sobre la ley agraria* (Study of Agrarian Law) (1794), was organized around epistemological concerns similar to those that informed the writings of Gándara and Arroyal.[25] Like Arroyal, Jovellanos thought that when it came to finding the cause(s) of Spain's decline, much prudence was needed, and he advised intellectuals to weigh models "with great care and circumspection so as not to put forth irresponsible theories on subjects in which errors have proven to have such a generalized and pernicious influence."[26] Yet for all his calls to prudence, Jovellanos sought to identify a single cause to account for all the ills of the nation.

Paradoxically, Jovellanos identified the cause of decline as precisely the hectic search for causes. Like Gándara, Jovellanos insisted that previous legislative efforts at reforming Spanish agriculture had contributed to making things worse. In fact, the desperate search for explanatory models to understand Spain's decline had led to the issuing of many contradictory laws. Reform for Jovellanos consisted in getting rid of these laws, which were "hurdles in the path to progress."[27] Jovellanos understood political economy to be a discipline charged not with designing ways to stimulate the economy but with removing obstacles. Such inclinations led him to embrace Adam Smith's doctrine of laissez-faire. The state for Jovellanos was an instrument of negative intervention, which should be entirely devoted to eradicating traditional entitlements and clearing up the tangle of old legislation. By insisting that the cause of decline was the poorly informed search for causes, Jovellanos sought to reduce the role of state intervention and reform to promoting education and building new infrastructure.

Juan Sempere y Guarinos was another leading eighteenth-century political economist who found wayward epistemological approaches to be the sole cause of Spanish decline. In his *Historia del luxo* (History of Luxury)

(1788), Sempere set out to prove that the continuous attempts on the part of intellectuals to identify luxury as the cause of Spain's decline was at the root of most problems. The fundamental problems of the nation, according to Sampere, consisted "in repeatedly coming up with false and inexact ideas on the most crucial points of legislation and politics; in often getting effects and causes confused; and [finally] in attributing to one [event] causes that belong to a very different one. From these factors stem our tendency to issue not only useless laws but also laws that often have the very opposite effect than the one legislators originally intended them to have."[28]

Sempere argued that the many laws introduced by the Habsburgs to curtail sumptuary consumption had destroyed national manufacturing. Seeking to eradicate the alleged moral decadence caused by luxury, lawmakers had passed laws that severely damaged local industries, promoted unemployment, vagrancy, and prostitution, and ultimately contributed to weakening the moral fiber of the nation. Moral philosophers had long blamed luxury for developments that in reality had completely different causes. Sempere, for example, sought to prove that the reported loosening of paternal authority and loss of female decorum among leading Spanish households had not been caused by profligate consumption that led to pampering elite children, as moralists had long held, but by the introduction of the *mayorazgo*, an institution that handed down the family patrimony to the first son, allocating inheritances regardless of the moral behavior of the children.[29] Thus, Sempere laid the blame for the decadence of Spain squarely on the shoulders of the elites, whose ignorance had blinded them into enacting "immoral and stupid" policies.[30]

Political Economy in the Masson de Morvilliers Affair

To shed light on the nature of the Spanish Enlightenment, many authors have focused on the Spanish debate triggered by the publication of Masson de Morvilliers's 1783 *Encyclopédie méthodique* article "Espagne." Morvilliers's disparaging views of Spain generated a groundswell of indignation and alternative views of how to reply his criticisms. In his magisterial *The Eighteenth-Century Revolution in Spain* (1958), Richard Herr maintained that these debates were emblematic of growing political and ideological tensions in the Iberian Peninsula in the wake of the French Revolution. Despite the various reforms introduced in the eighteenth century, particularly by Charles III, Spain had remained solidly united behind the monarch. The French Revolution and the arrival of a new king, Charles IV, changed the political landscape from consensus to growing polarization. Two parties, conservatives and liberals, emerged along rigid ideological lines. Herr used the Morvilliers affair to explore this tension. To typify the conservative side of the debate, Herr turned to the writings of Juan Pablo Forner, an intellectual who, according to Herr, mindlessly defended Spain while upholding traditional Christian values. In Herr's narrative, the writings of Luis Cañuelo, the editor of the weekly *El Censor* (The Censor), typified the liberal side of the debate. Cañuelo shared many of Morvilliers's dismal views of Spain, while urging immediate reform.[31]

This interpretation of the affair was challenged in 1967 by José Antonio Maravall. Forner, Maravall argued, was a proto-romantic, a patriot well acquainted with the idioms of the Enlightenment, not a conservative critic of secular modernity. Drawing on Maravall's insights and on the works of Antonio Mestre on Gregorio Mayans, whose scholarship Forner sought to emulate, the French Hispanicist François Lopez also presented Forner as a modern. According to Lopez, Forner had followed in the footsteps of Mayans and sought to recover the intellectual and religious traditions of Spanish humanism. His attack on Morvilliers and those who thought that Spain needed radical reforms was therefore no endorsement of ultramontane conservative ideologies.[32]

For all the virtues in Maravall's and Lopez's approach, I want to explore the debate from a different perspective. At the core of the controversy that followed the publication of Morvilliers's "Espagne," I argue, lay two views of political economy: one that advocated careful empirical historical research, and one that sought all-encompassing interpretations derived from first principles. These views were thus rooted in different epistemologies.

Forner sought to reply to Morvilliers, avoiding the path taken by previous patriotic defenses. Antonio José de Cavanilles and Carlo Denina, for example, had already answered the charges leveled by the French. Both replies, however, consisted of lists of prominent authors. This was a well-established genre. Patriots had long confronted European derision of the alleged Spanish intellectual backwardness by assembling bio-bibliographies. Nicolás Antonio's *Bibliotheca hispana nova* (1672) and *Bibliotheca hispana vetus* (1696), long lists of writings by accomplished Spanish authors, typified the genre. As European mockery grew over the course of the eighteenth century, this type of patriotic reference book multiplied. The political economist Juan Sempere Guarinos, who had identified the cause of Spanish decline in the secular tendency of the elites to misdiagnose the true causes of decline, also turned to the genre of bio-bibliography to demonstrate to

Europeans that Spaniards had made great intellectual strides under the Bourbons.[33] Juan Francisco Masdeu, Francisco Xavier Lampillas, and Juan Andrés, Spanish Jesuits in exile in the papal states who were confronted daily with charges by local intellectuals that Spanish bad taste had stunted the development of Italian literary traditions from Roman times to the early modern period, compiled lengthy bio-bibliographies from antiquity to the present, demonstrating the richness and variety of Spain's literary history.[34]

Forner chose a different line of defense. In his *Oración apologética por la España*, he sought to demonstrate that Spanish and French intellectual traditions had developed on the basis of radically different epistemologies. Turning the tables on Spain's foreign critics, Forner insisted that Spanish traditions were empirical, skeptical, and pragmatic. The French, on the other hand, were given to systematic, deductive, and ultimately speculative thought, most of it useless. Forner assigned Morvilliers squarely to the Cartesian intellectual style. Like Descartes, Morvilliers adored sweeping generalization, all-encompassing explanatory models, and ridiculous theories. French critics of the Spanish nation were all like Descartes and Morvilliers, engaged in endless speculative debates about the ultimate structure of things, from gravitation to the nature of magnetic and electric fluids. Forner pitted this style against the skeptical traditions of the Spanish Renaissance. Spaniards like Juan Luis Vives, he argued, understood that seeking to pinpoint the real causes of physical phenomena was pointless and thus turned to the study of ethics and the science of jurisprudence. Scholasticism, Forner insisted, was an Arab invention that had first spread to France, not Spain. It was when Spain betrayed its true skeptical self by embracing scholasticism that the nation had begun to decline.[35]

Forner championed a view of Spain as the cradle of the skeptical humanism that the likes of Vives and Francis Bacon had espoused. In his *Discurso sobre la historia de España* (Discourse on the History of Spain) (1796), Forner laid out an agenda for the discipline of political economy along the epistemological lines he had championed in *Oracion apologética*. Like any other history of Spain written in the eighteenth century, Forner sought to explain the causes of the nation's decline. But unlike his contemporaries, Forner thought that many details and factors remained to be elucidated through historical research. Forner went over the factors adduced by Raynal to account for Spain's decline, for example, including among others, the expulsion of Jews and Moriscos, hikes in taxation to finance chronic fiscal deficits by an overstretched empire, and persecution of new forms of knowledge. All these were plausible explanations, Forner said, but they were also tinged by ideological commitments that assumed that religious freedom, laissez-faire, and entrepreneurship caused growth. Raynal's commitment to ideology made him pick and choose the evidence. Forner had no time for this type of approach, for it was both reductive and dangerous. These models sought to put the "blame of decline on only one cause" and explained all symptoms deductively.[36] Such commitment to theories and ideologies promoted reforms whose effects were unpredictable: "either good or bad, useful or pernicious, wise or foolish."[37] The epistemological alternative to Raynal's narrative of Spanish decline, according to Forner, was meticulous empirical research. Forner urged intellectuals to visit archives and write empirically informed monographs on the causes of decline.

Forner practiced what he preached. In the very book in which he castigated ideologically informed theories of decline, he included a history of Spanish historiography, which rattled expectations and most likely irritated the king, because Forner claimed that the best scholarship had been produced under the Habsburgs, not the Bourbons. Quality had considerably declined under the latter. When it came to the history of historiography, narratives of early modern decline under Habsburg rule did not hold up to empirical scrutiny.

Forner's epistemology alarmed the editors of *El Censor*, a paper poking fun at all sorts of collective behavior published irregularly between 1781 and 1787 that was thrice shut down by the Crown.[38] *El Censor* in fact had a very serious intent. It sought to reform a Spanish culture perceived as corrupted by superstition and ignorance. For example, it took it upon itself to ridicule *villancicos*, or Spanish Christmas carols, whose lyrics, it asserted, promoted immorality rather than virtue.[39] To be sure, *El Censor* also made fun of the new Francophile culture of sociability and consumption. However, it forcefully denounced those who condemned luxury as the cause of Spain's decline, repeatedly running articles that endorsed commerce and trade as virtuous.[40] *El Censor* was the voice of a sector of the local intelligentsia who saw most foreign criticisms of Spain as reasonable and who wanted civic republican values to spread among the elites.

El Censor often ran articles by foreign travelers identifying the chief cause of Spanish decline as the excessive concentration of land in a few hands and the absence of a free market in land rents.[41] Although the editors seem to have believed that the true cause of Spain's decline lay in the lack of civic republican traditions among the elite, they did not dismiss foreign views out of hand. On the contrary, they welcomed foreign critiques as the only way to identify and correct the nation's problems.[42] *El Censor* had no patience with patriotic apologies for things Spanish and derided the compiling of bio-bibliographies, for they lulled the nation into believing that there was no need to change anything in light of such unparalleled intellectual accomplishments.[43]

El Censor harshly criticized Forner's *Discurso apologético* for having handed the enemies of reform a subterfuge to delay implementing change.[44] Forner's epistemology, it asserted, sanctioned political paralysis and ultimately conservative agendas. In the process of articulating a critique of Forner, *El Censor* offered an alternative view of political economy. If Forner construed the discipline as a branch of history, *El Censor* took it to be a branch of medicine. The metaphor of the body had long been applied to the study of society. Hobbes, for example, had compared a well-regulated hierarchical polity to the harmoniously well-integrated operation of the parts of the body.[45] But *El Censor* took this analogy one step further.

The Spanish body politic, *El Censor* argued, was ill. Sick bodies might have many symptoms, but these were all manifestations of a single disease. The medical metaphor allowed *El Censor* to articulate a view of political economy as a deductive science. Decline, like disease, should be explained by monocausal theories. It was the task of the critic to be the physician of the body politic, rooting out the cause of the malaise immediately. Delaying therapy was not an option.[46]

The debate sparked by the publication of Masson de Morvilliers's article suggests that political economy in Spain developed along two rather different paths. Fearful of a record of reforms that in the past had done more harm than good, and incensed by the mocking criticism of the French, writers like Forner insisted that political economy needed to avoid becoming Cartesian. Instead of seeking to bring disparate social phenomena into a single narrative of decline though deductive logic, these writers turned political economy into a historical discipline. Another group begged to differ. For them, Forner's historicism was simply a plot by conservatives to avoid reform. Drawing on the metaphor of the body politic, this group presented political economy as a deductive science meant to inform immediate social reform.

Conclusion

Eighteenth-century Spanish authors maintained an ambiguous and paradoxical relationship with foreigners. The Valencian Gregorio Mayans y Siscar promoted the concept of a Spanish Golden Age, and Cadalso condemned foreign observers as shallow and unreliable, while at the same time maintaining that Spain had historically been reinvigorated by cycles of foreign intervention. One is tempted to characterize the Spanish Enlightenment as nationalistic or even narrowly provincial, bent on denouncing foreign critics of things Spanish, for intellectuals were obsessed with setting the record straight and defending the nation. Their patriotism, however, was of a "cosmopolitan" kind. Intellectuals dismissed foreign critics precisely because the latter sought to present Spain as only marginally "European." Eighteenth-century authors furiously sought to assert their Europeanness, and in the process, they invented the Spanish Golden Age.

The evidence presented in this chapter shows that these authors were right. The debates over causation in political economy demonstrate the "Europeanness" of Spanish discourse. The ideas held by Forner, for example, anticipated Edmund Burke's in almost every respect. Like Forner, Burke linked his epistemological critique of the French philosophes to conservative theories of change and reform. Like Forner, Burke privileged historical and empirical research as the appropriate foundation of the human sciences and dismissed French speculative theories and systems as politically dangerous. Clearly, Forner and Burke belonged in the same cultural world.

CHAPTER 6

How Derivative Was Humboldt?
Microcosmic Narratives in Early Modern Spanish America
and the (Other) Origins of Humboldt's Ecological Sensibilities

In this chapter I offer an example of how a more generous understanding of the Atlantic world can yield new readings of the past, altering time-honored narratives. Alexander von Humboldt (1769–1859) has long been hailed as a founding father of the science of ecology and a genius. He has been credited with single-handedly creating a new discipline that relied on painstaking measurement to identify hitherto uncharted regularities in the mechanism of the planet as a whole. Humboldt demonstrated, for example, that individual species were part of larger plant communities, and that these communities were distributed geographically according to environmental variables such as elevation above sea level, temperature, and soil composition. After a lengthy visit to Spanish America (1799–1804), Humboldt went on to publish some thirty volumes on subjects ranging from botany to the political economy of Cuba and Mexico, consolidating his reputation as one of the leading scientists of the nineteenth century.[1]

Heroic narratives are today out of favor, and historians no longer portray Humboldt as a solitary genius. Janet Browne, for example, has situated Humboldt's ecological thinking within the larger history of the discipline of biogeography and shown that Humboldt drew on the ideas of contemporary German scholars such as Johann Forster (1729–98), Georg Forster (1754–94), and Karl Ludwig Willdenow (1765–1812). But for all her efforts to historicize Humboldt and to show his indebtedness to other Europeans, Browne remains wedded to the notion that exotic places are worth mentioning solely as backdrops to the exploits of European naturalists; she thus has no place in her study for the discourses and ideas circulating in Spanish America that could have influenced Humboldt.[2] In contrast, the British historian David Brading has shown that Humboldt's works on the political economy of Mexico and Cuba were possible only because he drew on the reflections of scholars in New Spain and on decades of data collection by colonial bureaucrats.[3] I have demonstrated elsewhere that Humboldt's antiquarian scholarship on Mesoamerican societies grew out of his encounter with rich empirical and interpretative traditions in Mexico.[4] Here I propose a new narrative for the history of biodistribution that treats the intellectual milieu Humboldt encountered in Spanish America seriously and that takes into account the discourses about space and nature formulated there during the seventeenth and eighteenth centuries.

The Euro-Creole Origins of the Science of Biodistribution

Several generations ago, the Catalan geographer Pablo Vila maintained that "geobotany was born of the encounter between two sages," namely, Humboldt and the late eighteenth-century Colombian naturalist Francisco José de Caldas (1768–1816). Pointing to the remarkable similarities between Humboldt's research program and that of Caldas, Vila forcefully argued for the "Euro-Creole" origins of the new science. Caldas was a self-taught naturalist and astronomer who, after having dazzled Humboldt, became one of the leading naturalists in the botanical expedition to New Granada of José Celestino Mutis (1732–1808), lionized by Humboldt for his fabulous collection of South American plants and natural history library. The powerful Mutis hired Caldas, who spent the following three years (1802 to 1805) traveling throughout the Ecuadorian Andes identifying and classifying varieties of trees and shrubs of the genus *Cinchona* (from whose bark quinine is derived) for his patron. Impressed by the results, Mutis called Caldas back to Bogotá, the capital of the viceroyalty of New Granada, to direct a brand-new astronomical observatory. Once in the city, Caldas also edited a weekly, the *Semanario de la Nueva Granada*, and became increasingly involved in the local patriotic societies that sought to change society through enlightened reforms. During the wars of independence triggered by the political vacuum in the colonies caused by Napoleon's invasion of Spain in 1808, Caldas joined the patriot forces as both ideologue and military engineer. Captured by Spanish troops in 1816, he was shot in the back by a firing squad, signaling the end of an entire generation of patriot naturalists.[5]

In 1801, when he encountered Humboldt in Colombia, according to Vila, Caldas was already charting the geographical distribution of plants in the northern Andes. By the time Humboldt published his 1805 *Essai sur la géographie des plantes* in France, Caldas had already produced several biogeographical maps of the northern Andes (1802), a memoir on the geographical distribution of plants near the equator (1803), and a study of distribution of *Cinchona* relative to height above sea level and temperature (1805).[6] These documents clearly indicate that Caldas was thinking about mapping bio-

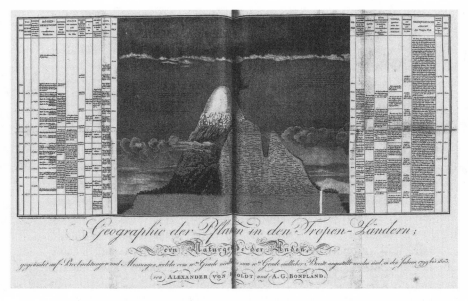

F IG . 6.1. Cross-section of the Andes from Alexander von Humboldt's *Essai sur la géographie des plantes* (Paris, 1805), showing correlations between plant communities, soil composition, and heights.

distribution in terms identical to those later made public by Humboldt. Vila therefore insists on what he calls the "Euro-Creole" origins of theories of biodistribution.

For all Vila's insights, the evidence clearly shows that Caldas learned cross-sectional mapping of heights from Humboldt, not the other way around. The Prussian naturalist found a lonely and self-taught Caldas botanizing in southern Colombia while making a living as an itinerant merchant. Humboldt was impressed by the creativity of this Andean "genius," for Caldas had built instruments from scratch, kept extraordinarily accurate astronomical observations, and invented a mathematical formula to calculate altitude by noting the temperature of boiling water at different heights. Caldas seems at first not to have been overly impressed by the Prussian and was skeptical of Humboldt's reliability as an observer: "Can we hope to get anything useful and knowledgeable from a man who would traverse our kingdom with so much haste [four to five months]? Isn't he going to broadcast prejudices and

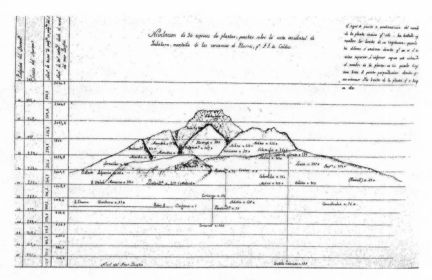

FIG. 6.2. Cross-section of the Andes by Francisco José de Caldas showing correlations between altitude and plant distribution, one of many graphics drawn by Caldas ca. 1802 that are strikingly similar to those later published by Humboldt. Caldas was inspired in this case by a cross-section Humboldt drew in October 1801. Source: Archivo del Real Jardín Botánico de Madrid (ARJBM), división III, signatura M-529. Published with the permission of the ARJBM. Many thanks to Daniela Bleichmar for helping me obtain this image.

false information to Europe, as almost all travelers do?" Caldas soon changed his tune, however, and looked forward to benefiting from his encounter with Humboldt (including the promise of a trip to Europe, which Humboldt later withdrew in haste when he met the handsome and aristocratic Marquis Carlos Montúfar in Quito, whom Humboldt subsequently took to Europe), saying: "I shall seek to learn and suck knowledge from this sage to gain some small measure of enlightenment and overcome [my] barbarism." Caldas obtained from friends a cross-sectional map of the Andean heights that Humboldt had completed sometime in October 1801, which inspired him to make one of his own (see figs. 6.1 and 6.2). Clearly, Caldas was the junior member in this so-called Euro-Creole partnership.[7]

Vila's assertion of the Euro-Creole origins of the science of biodistribution understandably bolstered patriotic pride in Colombia, but it left Eurocentric narratives in the history of science unchallenged.[8] For Vila's insights can be easily made to fit into diffusionist narratives of scientific discovery: Caldas emerges from more careful analysis simply as the precocious disciple of the learned European traveler Humboldt.

In this chapter, I seek to present an alternative narrative, one that focuses more on the origins of Humboldt's ideas about the Andes as a microcosm in which to test theories of biodistribution than on the origins of his cross-sectional graphics. Humboldt may have arrived in Spanish America with a scientific agenda already framed by the writings of Karl Willdenow and the Forsters. Once there, however, he encountered an intelligentsia obsessed with describing the rich local ecological variations. Humboldt learned to read the Andes as a natural laboratory for the study of the geography of plant communities in part because local Spanish American scholars had for decades (if not centuries) been developing this idea.

Paradise as Microcosm

Over the two hundred years before Humboldt's arrival, a tradition of natural history writing had arisen in Spanish America that considered the Andes a providentially designed region, seemingly endowed with all the climates of the world and thus potentially capable of producing any natural product. This tradition resulted from the meeting of Amerindian and European conceptions of space. On arrival, the Spaniards encountered civilizations in the Andes that from a European perspective exhibited curious patterns of settlement. Instead of relying primarily on markets to access commodities, Andeans sought to control resources by sending migrants to occupy various ecological niches. Andean communities were deployed in "vertical archipelagos," in the words of John Murra. Spaniards soon learned to take advantage of these peculiar spatial arrangements for purposes of labor mobilization and commercial agriculture.

The immensely rich diversity of ecological niches encountered by the Spaniards in the Andes prompted colonial scholars to associate the region with the biblical Eden, which, it was thought, had once contained all the fauna and flora of the Earth. In an effort to re-create this primitive paradise, naturalists in the Renaissance established botanical gardens. As John M. Prest has argued, the so-called discovery of America set off a vogue for collecting exotic plants in hopes of reviving paradise.

In the early modern period, mountains were second only to botanical gardens as substitute Edens. As late as the eighteenth century, Linnaeus imagined Eden as a very tall equatorial peak with a multitude of climates. The many microclimates of this mountain had once sustained all the fauna and flora of the world. As the oceans receded, however, species began to colonize distant geographical regions from the tropics to the Arctic, seeking environments that resembled the niches in paradise for which they had originally been designed. Linnaeus used the ancient construct of paradise as an equatorial mountain to explain biodistribution.[9]

Steep equatorial mountains with microclimates that reproduce those of the rest of the world were not merely bygone primeval spaces, however. They could be found in America. Christopher Columbus was perhaps the first to think that the lands he had just discovered had been the site of the Garden of Eden. Like his contemporaries, Columbus thought that paradise was at the top of an extremely tall mountain, the nipple of a breast-shaped peak that reached beyond the sublunar sphere. To be perfect, paradise had to transcend the laws of physics, and in classical cosmology heavenly matter in the celestial sphere was not subject to change. Only above the spheres of earth, water, air, and fire could the generation and transmutation of the elements be avoided.[10]

Although Spaniards did not discover peaks in the New World that rose high enough to be impervious to the laws of sublunar matter, they found an explanation in the Andes for why the torrid zone of America was temperate, even though it ought to be uninhabited owing to the scorching heat of the equatorial sun. Naturalists such as José de Acosta (1540–1600) held the Andes in awe when they discovered that climate was as much a function of elevation above sea level as of temperature. The equatorial mountain ranges Acosta encountered contained all the climates of the terrestrial sphere within relatively small vertical spaces. Spanish naturalists seeking to identify the meteorological mechanisms that kept the otherwise scorching tropics temperate seem to have ascribed paradisiacal properties to the Andes (and the New World generally).[11]

The first author to make this connection explicit was Antonio de León Pinelo (1590–1660), whose *Paraíso en el Nuevo Mundo*, written between 1645 and 1650, but not published until the mid twentieth century, sought to prove that the Garden of Eden had been located on the eastern slopes of the Andes. León Pinelo's work grew out of his dissatisfaction with all extant literature attempting to establish the site of Eden. Ancient learned consensus held that it had been situated somewhere in the Middle East or Asia. The new philological and geographical knowledge of the Renaissance gave novel twists to these age-old speculations. León Pinelo dismissed both new and old theories and argued that the correct reading of Genesis placed paradise in the Andes (fig. 6.3).[12]

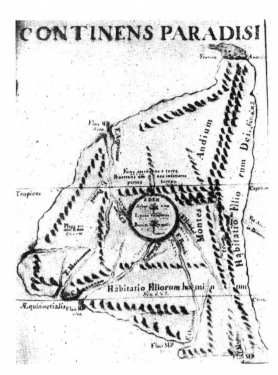

FIG. 6.3. Location of paradise in South America according to Antonio de León Pinelo, *El paraíso en el Nuevo Mundo* (1645–50), ed. Raul Porras Barrenechea (Lima: Imprenta Torres Aguirre, 1943), 1: 138.

To prove this, León Pinelo engaged in high-flying philological speculation. He demonstrated that the rivers Amazon, Magdalena, Orinoco, and Plate had the properties ascribed in Genesis 2:6–15 to the four rivers of paradise, namely the Gihon, Tigris/Heidekel, Euphrates/Perath, and Pishon. He showed that the reference in Genesis 3:24 to an angel with a flaming sword guarding the entrance to the garden was simply a metaphor for the Andean volcanoes surrounding Eden. He also argued that the tree of knowledge, the taste of whose fruit had caused the Fall of Man, had most likely been the South American granadilla, or passion fruit (*Passiflora edulis*), whose flowers and leaves resembled the instruments of Christ's passion on the cross (nails, sponge, lance, wounds, bindings, and crown of thorns), pointing to both original sin and its redemption (fig. 6.4). More important for my argument, León Pinelo maintained that of all places on earth, only the South American tropics near the Andes enjoyed the topographical and meteorological conditions that could have been home to a garden as temperate and as bountiful as Eden.

León Pinelo was skeptical that Eden could have been on top of a moun-

The Iesuites Figure of the Maracoc.

GRANADILLVS FRVTEX INDICVS
CHRISTI PASSIONIS IMAGO.

FIG. 6.4. Passionflower incorporating the instruments involved in Christ's passion, illustrated in John Parkinson's *Paradisi in sole paradisus terrestris* (London, 1629). Antonio de León Pinelo was clearly drawing upon a well-established tradition of theological scholarship concerning this flower, which begins with the publication of Simone Parlasca's *Il fiore della granadiglia, overo della Passione di Nostro Signore Giesù Christo, spiegato e lodato con discorsi e varie rime . . .* (Bologna: Bartolomeo Cocchi, 1609).

tain, for life in the Andes proved that the thin air of very high altitudes made breathing difficult. Nevertheless, he maintained that of all places in the world, only the Andes could have reached the middle region of the sphere of air, where corruption and the transformation of the elements were considerably retarded. In addition, the Andes helped him explain how a place on the equator, which should have been rendered uninhabitable by the scorching heat of the sun, was in fact the most temperate environment on earth. Andean heights offset the tropical position of Peru on the terrestrial sphere, yielding a perfect meteorological balance.

Once he had shown that Eden on the equator was not an oxymoron, León Pinelo set out to demonstrate that the natural history of Peru was sufficiently rich to make his case. His catalogue of local fauna and flora, however, was somewhat atypical, because he offered only a list of wonders (see, e.g., fig. 6.5). Forced to offer reliable criteria to measure the organic capacity of the terrain, León Pinelo turned to a description of curiosities, believing that the more wonders brought forth by the land, the more likely it had once been the location of Eden. It is not clear how León Pinelo drew the

Chapter 6

F I G . 6.5. A monster in Peru from Antonio de León Pinelo, *El paraíso en el Nuevo Mundo* (1645–50), ed. Raul Porras Barrenechea (Lima: Imprenta Torres Aguirre, 1943), 2: 116.

connection between natural wonders and the sacredness of the place, but early modern scholars thought that God best revealed his omnipotence and artistry through nature's play rather than through its regularities.[13] Judging by the sheer size of León Pinelo's catalogue of local wonders (it took up at least one-third of his treatise), Peru far surpassed any competitor.

León Pinelo also suggested that the abundance of microclimates in the Andes was the cause of the wealth of wonders he had catalogued. The Andes allowed him to explain why Peru was so bountiful. León Pinelo identified three habitats in the Andes, each distinctively rich in its own way: the low-lying areas of the coast and the Amazon, the middle ground, or *llanos*, and the high-altitude sierras. These multiple ecological niches rendered the area particularly productive, for as a crop withered in one niche, it flourished in another. More remarkable was the fact that its many microclimates made Peru hospitable to all crops and products. Whereas the plants of America were not easily acclimatized in Europe, all European crops yielded immense harvests in Peru.[14]

The Political Economy of Paradise

León Pinelo's forceful patriotic argument in his natural history is not an economic discourse; he was chiefly concerned with cataloging wonders and curiosities, not with investigating the Andes as a source of wealth. It fell to eighteenth-century intellectuals to undertake this task. To capture this transition, we may consider similar developments in Europe. Lisbet Koerner has shown that Linnaeus's taxonomy and natural history were intimately linked to "cameralist" (statist) discourses seeking to transform Sweden into a self-sufficient economy. Linnaeus sent students abroad to collect flora in the hope of weaning Sweden from its dependence on imports. According to this utopian view, through careful acclimatization of plants in botanical gardens, naturalists would provide all the raw materials needed for the kingdom of Sweden to become self-sufficient. Given Linnaeus's views on biodiversity, the effort consisted not merely in reproducing Eden but in making it economically viable.

Spanish Americans who lived in the Andes did not have to send naturalists abroad in search of sources of wealth. They simply turned to the microcosm next door. Unlike Linnaeus, moreover, Spanish American intellectuals in the viceroyalties of New Granada and Peru did not seek to make their countries self-sufficient, but rather to turn them into commercial emporiums based on the microcosmic ecological attributes of the Andes, supplying the consumers of the world with all they needed. A flurry of utopian debates on how to harness the untapped wealth of the Andes greeted Humboldt on his arrival. To understand these debates, we need first to understand the institutional and cultural context in which they took place.

Like their British-American cousins, who felt entitled to their "English freedoms," Spanish American settlers had enjoyed unparalleled degrees of autonomy and self-rule until the post–Seven Years' War reforms. Spanish American societies were viceroyalties, not colonies, autonomous polities in the loosely held composite that was the Iberian Catholic monarchy. These "kingdoms" (hierarchical polities organized on the principles of socioracial estates and corporate privileges) enjoyed numerous forms of local political representation (from city councils to cathedral chapters), which came under attack with the Bourbon reforms.[15]

Determined to transform these viceroyalties into colonies, the Spanish Bourbons turned to the new sciences. The Spanish empire had long been losing territories along with status and prestige in the New World to other European powers. Some Spanish intellectuals maintained that the loss of territories began with losses in the struggle over naming, surveying, and remembering. The writing of histories of "discovery" and colonization and the launching of cartographic and botanical expeditions therefore became priorities for the state, and many such expeditions visited the New World. Naturalists sought to benefit the economy by identifying new products (dyes, spices, woods, gums, pharmaceuticals) or alternatives to already profitable staples from Asia. Spanish botanical expeditions to the Andes, for example, put a premium on finding species of cloves and cinnamon to challenge the British and Dutch monopolies in the East Indies.[16] The logic behind sending botanical expeditions to the New World was best expressed in 1777 by the architect of these policies, Casimiro Gómez Ortega (1740–1818), who assured José de Gálvez (1720–87), minister of the Indies, that "twelve naturalists . . . spread over our possessions will produce as result of their pilgrimages a profit incomparably greater than could an army of 100,000 strong fighting to add a few provinces to the Spanish empire."[17]

One of these expeditions was organized to survey and give names to the resources of New Granada, a territory that had recently been transformed by the Spanish Bourbons into a viceroyalty, with its administrative center in Bogotá, in the hopes of putting an end to the immensely profitable British and Dutch illegal trade off the coast of Venezuela. José Celestino Mutis was the head of the expedition, and his ideas typify the spirit of the enterprise; they also capture how quickly notions of the Andes as a microcosm were grafted onto the original Spanish expeditionary project.

Mutis arrived in Bogotá in 1761 as part of the viceroy's entourage. He quickly set out to explore the land, and when he heard about the official campaign to send botanical expeditions to the New World, he requested that his efforts be acknowledged. In 1783, he found himself in charge of the so-called Botanical Expedition of the New Kingdom of Granada.[18] Spanish merchants had long benefited from their monopoly on quinine. In a century in which fevers were at the center of medical thought in Europe, the virtues of quinine as a febrifuge made the trade in it extremely profitable. Quinine came from the bark of a tree found in a small area of Loja on the eastern slopes of the equatorial Andes. Mutis was determined, however, to find quinine-producing cinchona trees on the Andean slopes of Colombia as well. Mutis eventually found new species in Colombia, although they were different from the trees of Loja. He simply assumed that similar areas (elevation above sea level, temperature, distance from the equator) should produce similar trees.[19]

Many of the efforts of the expedition under Mutis were fueled by the assumptions that similar environments engendered similar botanical species, and that the Andes were a treasure trove of microclimates. Thus, in 1785, Mutis claimed to have found a substitute for Asian tea in Colombia and set out to convince the Spanish authorities that his Colombian product was as good if not better than the tea Europeans had been importing from China. Behind these efforts lay the idea that the Colombian Andes were providentially furnished with microclimates capable of supplying the world with any product. "Countless are the natural productions with which Divine Providence has endowed this New Kingdom of Granada," Mutis argued in a letter sent to the Spanish minister of state, José Moñino y Redondo, count of Floridablanca (1728–1808), whose recent illness, Mutis suggested, the new Colombian tea would likely cure. The striking organic potential of New Granada, Mutis maintained, was due to the fact that this kingdom "was like a center of the Americas in which similar or equivalent productions to those found in the immense space of the Old and New Worlds have been gathered."[20]

The members of the expedition led by Mutis did their utmost to spread the news of the fantastic economic potential of New Granada. For example, around 1790, the Creole lawyer Pedro Fermín de Vargas (1791–1830), a member of the first phase of Mutis's expedition (1783–91), portrayed New Granada as a land of unparalleled commercial potential. According to Vargas, the viceroyalty enjoyed an exceptionally good geographical location, where it was possible "to find almost all the climates of the globe." Colombia was a microcosm owing to the multitude of ecological niches created by the Andes and to the endless agricultural cycle of its equatorial climate. It was also a potential economic leader of the world. If an enlightened ruler were to build roads and to protect and increase the population to accelerate the "circulatory rhythms" of the country, Colombia, according to Vargas, would be poised to supply the world with cinnamon, cloves, tea, betel pepper leaves (a stimulant that is chewed in Southeast Asia and that could have been replaced with coca leaves), and indigo of even better quality than the Asian equivalents. The coastal plains of Cartagena and Santa Marta alone would provide the cotton needed by all factories of the world.[21]

Other members of the expedition used the pages of the *Papel Periódico de la Ciudad de Santafé de Bogotá* (1791–97), a periodical created by fiat of the viceroy in order to stimulate a colonial public sphere and to spread the optimistic message that the key to the future prosperity of the viceroyalty lay in its microcosmic qualities. Francisco Antonio Zea (1770–1822), future director of the Royal Botanical Garden in Madrid, led the charge for this message in 1790. Under the pseudonym Hebephilo, Zea called on the youth of Bogotá to become citizen patriots, interested solely in the greater good of society. Such virtuous citizens armed with the tools of the new sciences would one day see New Granada become a trade emporium, for "here Nature has shown herself in all her magnificence; here [she] has revealed even to the blind and the ignorant the bright pageant of her marvels."[22] A few days later, a reader of the *Papel Periódico* made Zea's views explicit by clarifying the reasons to be hopeful. "Nueva Granada," the reader argued, "is surrounded by the most beautiful and most diverse variety of climates, which are located at very little distance from one another." Moreover the land could produce any type of natural commodity in the world, including balms, gums, medicinal plants, cotton, wheat, legumes, fruits, cattle hides, wool, birds, precious stones, reptiles, metals, coveted mercury for amalgamation of gold and silver, and even East Indian cinnamon and cloves. Only idleness, ignorance, and lack of civic virtue, the reader posited, could keep the new kingdom of Granada from fulfilling its unlimited economic potential.[23]

It is not surprising, therefore, that when Humboldt arrived in New Granada with rigorous new techniques to measure and chart biodistribution, Caldas embraced them almost overnight, producing studies and maps before Humboldt had the chance to publish his own. It was this prompt embrace that led Vila to suggest mistakenly that biodistribution was an idea created by both Humboldt and Caldas. In fact, Caldas's charts and maps were simply spatial representations of much older ideas.

Like Mutis, Vargas, and Zea, Caldas was deeply committed to the notion that New Granada was a microcosm, providentially designed to enjoy unlimited economic potential. "From the bosom of New Granada," Caldas insisted, "all the perfumes of Asia, African ivory, European industrial commodities, northern furs, whales from the South Sea, [in short], everything produced on the surface of our world [can be obtained]."[24] The microcosmic attributes of the Andes prompted Caldas to portray New Granada as a natural laboratory in which to study correlations between behavior, race, and climate.[25] In addition to being a microcosm, New Granada was geographically designed to be a trade emporium, a new Tyre or Alexandria. It was located at the center of the world and equipped with navigable rivers to carry staples from the interior to the coast, as well as ports facing both the Atlantic and the Pacific. "Nueva Granada," Caldas maintained, "appears destined for greatness by its geographical position for universal commerce."[26] This type of logic led another member of Mutis's expedition, the naturalist Jorge Tadeo Lozano (1771–1816), compiler of an as yet unpublished "Fauna of Cundinamarca," to predict in 1806 that his *patria* was poised to become in "a few centuries a vast empire that . . . will equal the most powerful in Europe."[27]

These ideas also surfaced in Peru, another place that witnessed botanical studies sponsored by the Spanish Crown. Hipólito Unanue (1755–1833), editor of the Lima periodical *El Mercurio Peruano* (1791–95) and a physician committed to reform, typifies the scholars in Peru who gave León Pinelo's old ideas a new twist. Like Caldas, Unanue thought that his *patria* was destined to become a trade emporium. In addition to the microcosmic qualities of the Andes, Unanue focused on the physical features of the land, pointing to Peru's as yet unfulfilled potential. "It seems," Unanue argued, "that after having created the deserts of Africa, the fragrant and lush forests of Asia, and the temperate and cold climates of Europe, God made an effort to bring together in Peru all the productions he had dispersed in the other three continents. In this manner, God has sought to create [in Peru] a temple for himself worthy of his immensity, [a temple] majestically surrounded by all the treasures hidden in this kingdom."[28] Peru was, in short, "the most magnificent work Nature has ever created upon the Earth."[29] God had revealed a predilection for Peru in the subtleties of its physical structure. For instance, Peru had been chosen by God to keep the balance of the planet. The massive weight of the Andean mountains was responsible for tilting the Earth's axis and thus for the very existence of Europe, which otherwise would have remained under water.[30] Like Mutis, Unanue speculated that certain local products were suitable substitutes for popular products currently monopolized by Spain's European rivals. Coca, whose sharp, acrid particles stimulated circulation and digestion, could one day replace tea and coffee in the global economy.[31] Enjoying such unparalleled physical properties, Peru was poised to supply the world with all it needed.

Curiously, these ideas about the microcosm found audiences throughout the Spanish American lands, including the mountainous areas of Mexico and flat plateaus such as those of Buenos Aires. For example, Creoles in Buenos Aires, who had long called the pampas a "desert," useful only for wild-cattle grazing, also saw the viceroyalty of La Plata as a microcosm. In their imaginations, La Plata was a land of multiple ecological niches poised, like ancient Tyre, to be "the center of all the commercial circulation of the world" and, like ancient Alexandria, "a port connecting the East and the West." In his 1799 inaugural address to the Nautical Academy, financed by the Consulado of Buenos Aires to train qualified sailors for a merchant navy, Pedro Antonio Cerviño (d. 1816) called attention to the Argentine capital's privileged central position in the world. "Our location [on the globe] is a most felicitous one," Cerviño maintained, "[because] North America, Europe, Asia, and the Pacific Ocean are all equidistant to us. This marvelous location assures us an immense commercial traffic. [We] shall become the warehouse of the universe."[32] In 1801, Francisco Antonio Caballe, editor of the short-lived Buenos Aires *Telégrafo Mercantil*, depicted the viceroyalty of La Plata as a land capable of supplying the world with hides, tallow, wheat, cocoa, quinine, indigo, copper, henequen, and "all sorts of resins and drugs, not to mention precious and abundant gold and silver . . . [as well as the equally precious] saltpeter, pearls, and seashells that can be found in the spacious Chaco." Caballe concluded, "without recourse to hyperbole," that in all the world there was "no other land as rich, holding such a variety of products . . . and [thus] as suitable for establishing strong and powerful commercial institutions" as the viceroyalty of La Plata.[33] By 1802, it had become a truism among Creoles that their *patria* was "like a sea, [in which] we lose ourselves in the horizon . . . a land of wondrous mountains with the best wood in the universe," a land "located [right] at the center of the commercial world and deliciously situated on the banks of a mighty river"—in short, a land "with the greatest productive power in the globe."[34]

Mexico also found an avid audience for these kinds of microcosmic narratives. Juan Manuel de San Vicente, for example, argued in 1768 that Mexico, like Babylon, was "the world writ small (an *epitome*)"; its markets demonstrated the abundance of "this second terrestrial paradise."[35] The physician Juan Manuel Venegas in 1788 offered cures for all sorts of diseases with prescriptions based on Mexican plants. For Venegas, New Spain was "the purse of Omnipotence; an Eden capable of providing Europe not only with precious metals but also with many of the noblest vegetables, roots, woods, fruits, gums, and balms."[36] And José Mariano Moziño (1757–1820), one of many Creole naturalists intent on writing a materia medica based on local plants and Nahua herbal lore, was convinced that "every single medicinal substance [in the world], with the exception of some three or four, can be abundantly supplied by our land. [Mexico] produces, if not the same medicinal botanical species, others that are of equivalent or perhaps of superior efficacy."[37]

Conclusion

Natural history and botany played significant economic and ideological roles in the early modern world. In the first phase of colonization of the New World, most Europeans single-mindedly pursued mineral riches. From Hernán Cortés to Walter Raleigh, conquistadors and explorers saw the New World both as an obstacle on the way to Asia and as an endless source of gold and silver. This was all to change by the seventeenth century. The emergence of a fledgling mass-consumption society in northern Europe set off a plantation boom on the Atlantic shores of the Americas, based on the ruthless exploitation of slave and indentured labor. The new wealth of the Americas suddenly turned "green." Growing, harvesting, and distributing sugar, tobacco, coffee, indigo, rice, and quinine, to name only a few food staples and drugs, became sources of fabulous new wealth for both governments and merchants.[38] Even though the Iberian Catholic monarchy had long profited from the botanical trade (e.g., cochineal, sarsaparilla, quinine, vanilla, cocoa), Spaniards proved slow to adjust to this new era of plantations and economies of scale: under Philip III (r. 1598–1621) and Philip IV (r. 1621–65), the Portuguese footholds in the Indian Ocean and the China Seas were picked off one by one by the Dutch, who transformed the loose multiethnic, maritime trading networks of Southeast Asia into monopolies to control the production and distribution of nutmeg, cloves, and pepper. The Catholic monarchy could not stem the northern European barrage in the New World either. It was only in the late eighteenth century that, under the command of a new Bourbon dynasty, the Spanish monarchy began to compete in the new global agricultural-botanical markets. By then it was simply too late. The possessions of Spain in the New World had already become autonomous viceroyalties with traditions of self-rule and historical identities resistant to easy colonial subordination. The well-intentioned plans of botanists to use the New World to grow cloves, cinnamon, and other spices to break the Dutch and British monopolies proved to be a mirage.

But if the great Bourbon botanical plans never materialized, the cultural transformation they brought about was profound. In societies that had long considered themselves independent for all practical purposes, the new botany was cultural capital, giving rise to discourses on the untapped, providential economic potential of each of Spain's American viceroyalties. Projects designed to turn the viceroyalties into subordinate appendages of a new, revitalized modern empire unwittingly offered ideological tools that allowed those communities to see themselves as central to the world. Narratives originally deployed to prove that America had been the location of the biblical Eden were suddenly redeployed under duress in order to reimagine the future of the viceroyalties.

The encounter with the Andes and the long-standing tradition of thinking about the American viceroyalties as Edenic microcosms could not have failed to impress Humboldt. His trip to the New World was not planned in advance: he had actually been trying to go to Egypt. He did not voyage to South America deliberately to prove the Forsters' and Willdenow's speculations on biodistribution true; his encounter with the Andes was serendipitous. Prompted by the ceaseless rhetoric about the microcosmic virtues of the Andes, Humboldt began to think of these mountains as a laboratory for testing theories of biodistribution. Historians have managed to write histories of biogeography without acknowledging that crucial components of Humboldt's ideas did not emerge in Europe. Humboldt arrived in a Spanish America humming with discourses of nature, in which each *patria* was depicted as providentially destined to become a trade emporium. Humboldt learned to read the Andes as a natural laboratory for studying the geography of plant communities only because local scholars had been toying with this idea for years.

CHAPTER 7

Landscapes and Identities
Mexico, 1850–1900

Sometime in 1865, Manuel Orozco y Berra (1816–81) wrote an indignant letter to the French minister of education about France's plan to map Mexico. With Maximilian of Austria on the throne as proxy ruler, the French military occupiers now set out to chart their new protectorate. For Orozco y Berra, an engineer long involved in Mexican government mapping initiatives, the current project was particularly galling because the French assumed that no one had previously attempted to map the republic.

To prove his point to the French minister, Orozco y Berra sent books, maps, and a memoir in which he complained bitterly about French prejudices. The French, he argued, regarded Mexico as a land of exotic "Indians," absolutely lacking in traditions of scholarship. They assumed that Mexico was a "wild," uncharted territory whose mapping had to be done from scratch. Reflecting on this episode many years after the occupiers had been driven out, Orozco y Berra acknowledged that the French had in fact advanced geography as a science in Mexico, but contended that their contributions had been limited to charting peripheral territories. Their maps of the long-settled "civilized" core had turned out to be unacknowledged copies of previous maps by Mexican engineers. Orozco y Berra also wrote a thick volume on the history of geography in Mexico that demonstrated the richness of Mexican scholarship, not only during colonial times but also during the national period, an era characterized by economic decline, civil wars, extraordinary territorial losses, and humiliating foreign occupations. Orozco y Berra went on to describe in painstaking detail the many nineteenth-century geographical and cartographic expeditions put together by the Mexican federal government and by state authorities from Chihuahua to Chiapas.[1]

This little-known episode in the history of nineteenth-century Mexico has not lost its relevance. To this day, Western scholars still manage to write off entire chapters of the Latin American experience based on the same assumptions that once moved the French to ignore centuries of cartographic and geographic scholarship in Mexico. Take, for example, the case of landscape painting, a subject intimately linked to Orozco y Berra's own concerns. Were one to trust current writings in English (or French, German, or Italian for that matter) on this subject, one would be forced to conclude that no landscape tradition worth studying ever emerged in the region. Apart from a handful of essays in English on the great nineteenth-century Mexican landscape painter José María Velasco (1840–1912), mostly written by Mexican scholars, there is not much to go on.[2] From this paucity of evidence, the unsuspecting public must conclude that the traditional narratives of Western art that are served up in museums across the globe are the complete story. Mexican scholars have written a great deal about Velasco and nineteenth-century landscape painting more generally, however, and in the following pages I draw extensively on their researches.[3]

This chapter does not seek to do for landscape painting in Latin America what Orozco y Berra did for the history of cartography and geography in Mexico. A cultural history of this genre in Latin America has unfortunately yet to be written. Numerous useful exhibition catalogues and detailed biographies of leading nineteenth-century Latin American landscape painters have appeared, but with remarkable exceptions, much of this literature either contents itself with locating the paintings (a significant achievement, given that many are in private collections) or falls short in interpretation.[4] My efforts are more modest. I simply seek to put Velasco's work in the wider context of nineteenth-century Mexican aesthetic and scientific discourse, locating it in the larger tradition of representations of nature and nation building. By focusing on paintings, poems, illustrations, and scientific treatises, this chapter seeks literally to bring some color, variety, and density back to the telling of the Mexican past, which has unfortunately been saddled with historical narratives that, like bad movies, are riddled with predictable plots. I also seek to put the cultural history of sensibilities about nature among Mexico's literate groups in a wider global context, drawing attention to similarities and differences, particularly with respect to developments in the United States.

Landscape and National Identities

Landscape paintings are privileged windows through which to peek into processes of nation building. Images of dunes, canals, castles, windmills, and cattle, for example, were intimately related to the efflorescence of patriotism in the seventeenth-century Netherlands. A population that came to develop a sense of collective identity only in the wake of the wars of religion with Spain, the Dutch almost single-handedly invented the genre. Dozens of painters dwelled on the providential signs of bounty in the land. By subtly lengthening its historical roots, landscape painters also gave the rapidly changing new nation—its territory transformed by international trade and increasingly crisscrossed with dikes and canals—a patina of longevity.[5] Other regions of the world also witnessed the rise of landscape painting as part of the search for new collective identities. The art of Caspar David Friedrich (1774–1840) typifies an entire generation of German romantics who sought to symbolize national resistance to Napoleon's invasions through representations of trees and forests.[6] Australian nineteenth-century painters, who like the rest of the population huddled in cities along the coast, imagined their new homeland as either an "empty" wilderness free for the taking (the bush) or a pastoral republic of hardy farmers transforming the rugged interior.[7] Similar tropes informed the aesthetic sensibilities of several generations of U.S. artists, whose works have collectively come to be misleadingly associated with the Hudson River School of painting.[8]

Less well known, to be sure, are the paintings of those Spanish artists who in the course of the nineteenth century grappled with similar identity questions. The romantic landscapes of Jenaro Pérez Villaamil (1807–54) squarely located the essence of the nation in the rugged Cantabrian Mountains of the north, where the medieval Catholic monarchies had kept the invading Arabs at bay. These idealized mountains helped underwrite liberal narratives of ancient gothic constitutions limiting the absolute power of the monarchy. By the end of the century, Pérez Villaamil's mountains give way to the somber Castilian landscapes of Aureliano Beruete (1845–1912), whose brownish palette and arid plateaus speak to the nostalgia and disillusionment of a generation who haplessly witnessed the decline of Spain as an imperial power.[9] Beruete's landscapes reflect the pessimism of a fin de siècle intelligentsia who found Spain to be closer to Africa than to Europe: a nation handicapped by an environment that had fostered a culture of idleness, a nation that to "regenerate" needed large works of engineering to irrigate the land.[10] Ambiguously, this same arid landscape spoke to these intellectuals of an environment capable of spawning character types remarkable for their endurance and honor, types that had caused the nation to achieve its past glories in the first place.[11] Far less invested in reforming the nation through the renewal of the Castilian core, Catalan and Valencian painters struck out on their own, creating idealized pastoral landscapes of a verdant Mediterranean world.[12]

Mexico: History in the Landscape

Nineteenth-century representations of the landscape in Mexico also reflect the anxieties and desires of intellectuals busy crafting a nation beleaguered by civil unrest and economic decline. Like their Australian, U.S., and European peers, Mexican artists sought to find the nation in representations of nature. A central and peculiar element in the patriotic appropriation of the landscape in Mexico, which will help us make sense of Velasco's paintings, was the development of aesthetic sensibilities that found it hard to pry nature apart from history. Mexican intellectuals described, painted, and relished landscapes largely because the latter were linked to local historical narratives.

This, to be sure, was part of a venerable European tradition. In most of seventeenth- and eighteenth-century Europe, classical and neoclassical landscape painting emerged as a minor pictorial genre in art academies, overshadowed by religious and historical genres. Artists such as Claude Lorrain (1600–1682) and Nicolas Poussin (1594–1665) gave landscapes prestige by making them historical. Picturesque scenes sought to evoke pastoral narratives from antiquity. Drawing upon Virgil, Horace, and Ovid, painters represented shepherds and nymphs cavorting in the Roman Campagna surrounded by classical ruins. Such was the grip of this tradition that academic painters in France still chose Italianate scenes over "national" ones well into the nineteenth century. The Barbizon School largely owes its prominence in modern histories of art to the fact that its members ceased to make Italy their focus and dropped all references to classical and biblical historical narratives.[13]

There is no doubt that neoclassical sensibilities also informed the way Mexican intellectuals represented landscapes in the nineteenth century. Many leading poets ensconced in the Mexican highlands remained committed neoclassicists, given to imagining classical and biblical scenes with Italianate backgrounds.[14] Moreover, painting in Mexico was firmly regulated throughout most of the century by the Academy of San Carlos, originally founded by Charles III in 1781 to spread good neoclassical taste, whose doors remained uninterruptedly open throughout the nineteenth century, despite the secular turmoil that engulfed the body politic. The academy privileged history painting over any other genre, and landscapes were tolerated only as backgrounds for biblical, classical, or pastoral scenes.

Despite the influence of neoclassical taste, or perhaps due to it, intellectuals also linked local landscapes to local history. The growing awareness among nineteenth-century Mexicans of the beauties of Chapultepec, a forested hill near the capital, typifies the development of this new aesthetic sensibility. Chapultepec had once been a sacred forest, a courtly retreat for such great Aculhua (Texcocan) and Mexica (Aztec) rulers as Nezahualcoyotl and Moctezuma. This was all well known to any local schoolchild, but by the early nineteenth century, it took on new meanings. In 1839, for example, Ignacio Rodríguez Galván (1816–42) published his poem "Profecía de Guatimoc" (Cuauhtemoc's prophecy) in Mexico City, which typifies the new aesthetic sensibilities of the day. Abandoned by friends, spurned by the woman he loves, bereft by the recent death of his mother, and having committed his elderly father to an institution for the infirm and the poor, the poet decries humanity's evil. A ghost, the spirit of Cuauhtemoc, the last Aztec emperor, who was imprisoned and tortured by Cortés, suddenly shows up and engages the beleaguered poet in dispirited dialogue. Both the poet and the ghost conclude that the world is ruled by corrupt, mean-spirited elites; that Mexico is in the hands of effeminate, selfish regional bosses who will put up no resistance to invading foreign armies (an eerily prophetic statement); and that Amerindians have given up all hope of ever recovering the status, wealth, and power they once enjoyed. Cuauhtemoc predicts, however, that in the end justice will be served: the powerful will be humbled and the virtuous rewarded. What is remarkable about this poem is that it unfolds at night in Chapultepec as the poet wanders searching for romantic relief in the solitude of the forest. The physical setting of ancient trees and water springs elicits a poetic catharsis. It is clear from the poem that the forest is enormously pleasing. But such aesthetic appreciation cannot be dissociated from the past. The forest literally summons back history in the form of a ghost, Cuauhtemoc. Rodríguez Galván's gothic sensibilities did not last long, but his historically informed landscape aesthetics endured.

From 1843 to 1845, the reading public in the capital and the nation at large grew fond of *Museo Mexicano o Miscelanea pintoresca de amenidades curiosas e instructivas*, a periodical designed to introduce audiences to the marvelous heterogeneity of Mexican life and nature. In volume after volume, readers enjoyed images and read descriptions of local landscapes. What is striking about the articles and illustrations is that they invariably seek to connect places to national historical narratives. J. Soto, a reporter traveling through Veracruz, for example, proved unable to come up with a better picturesque view of the land than the Puente Nacional, a bridge in the road linking the city of Veracruz to Perote on the Gulf Coast, a piece of late colonial Bourbon engineering (fig. 7.1). The author of the article uses a view of the bridge to deliver a lesson on the battles for independence fought

FIG. 7.1. *Puente Nacional*, in J. Soto, "Panorama de México: El Puente Nacional," *Museo Mexicano o Miscelanea pintoresca de amenidades curiosas e instructivas*, vol. 2 (1843), 256–59. This engraving typifies a sensibility toward nature and the landscape in nineteenth-century Mexico: aesthetic perceptions are mediated through history and memory. In this case, a bridge and its vicinity where battles in the wars of independence were fought becomes a metonym for Veracruz. The landscape cannot be dissociated from the historically meaningful events that it has witnessed. The bridge, on the other hand, can be seen as a gesture warning foreigners that Mexico is not a romantic wasteland, the empty, uncivilized space images of which painters were sending back to Europe and the United States in the wake of Alexander von Humboldt.

in its vicinity. The forested hill of Chapultepec also figured among the landscapes and picturesque scenes discussed in *Museo Mexicano*, again viewed through an aesthetic prism colored by history (fig. 7.2). Guillermo Prieto, the author of the article that accompanies the etching, waxes lyrical about a forest that is a soothing maternal bosom, about the beauties of its springs and ancient trees, and about the sublime aerial views of the valley of Mexico to be enjoyed from the castle at the top of it. Several pages about the history of the forest and of the castle/fortification built in the late colonial period, however, temper such unrestrained romantic lyricism.

Multiple aesthetic readings of Chapultepec surfaced over the course of the century, all permeated by history. In *México y sus alrededores* (1855–56), a collection of engravings of Mexico City and its vicinity edited by José Antonio

FIG. 7.2. *Chapoltepec*, in Guillermo Prieto, "Chapoltepec," *Museo Mexicano o Miscelanea pintoresca de amenidades curiosas e instructivas*, vol. 3 (1844), 212–16. The forest of Chapultepec was perhaps the most aesthetically pleasing of all landscapes Mexicans came to appreciate in the nineteenth century, largely because it was steeped in history. It contained hundreds of sacred *ahuehuetes*, grounding Mexico in its precolonial past, and was the site of heroic resistance against the invading troops of the United States.

Decaen, Chapultepec is portrayed as an ancient primeval forest, a piece of the original Mexico left unspoiled by the European conquest. "Only you, great forest," Luis de la Roca proclaimed in a bout of lyrical frenzy, "have survived so much devastation and so many ruins!"[15] In typical orientalist fashion, de la Roca then proceeds vicariously to enjoy scenes of Moctezuma's seraglio as he walks through the forest. If, for nineteenth-century German poets, the forest became the cradle and crucible of the nation's virtues, for the Mexican intelligentsia, Chapultepec evoked the tragedies of conquest and the gothic, ghostly presence of a grandiose yet spent past. Chapultepec also became a romantic retreat where intellectuals sought to gain insight into the meanings of solitude. After the U.S.-Mexican War, the forest became even more tightly linked to patriotic historical narratives, for Mexicans came to associate it with the martyrdom of *los niños heroes* (the boy heroes), a handful of courageous teenagers who resisted the U.S. military assault on the hill's fortifications to the death. Dozens of landscape paintings of Chapultepec appeared over

FIG. 7.3. Casimiro Castro, *La glorieta en el interior del bosque de Chapultepec*, in Luis de la Roca, "El bosque de Chapultepec," in *México y sus alrededores* (1855–56), 25.

the course of the century, by such male and female artists as Pelegrín Clavé (1811–80), Pedro Gualdi (ca. 1847), Cleofás Almanza (1850–1915), Germán Gedovius (1867–1937), José Inés Tovilla, Casimiro Castro (1855–56), Dolores Soto de Barona (1869–1964), Guadalupe Velasco (fl. late nineteenth century), and José María Velasco (figs. 7.3 and 7.4).

The *ahuehuete*, Tule tree, or Mexican cypress (*Taxodium mucronatum*) was another staple of a nineteenth-century Mexican aesthetic imagination thoroughly disciplined by history. Writers and painters were dazzled by Chapultepec largely because it harbored one of the largest stands in Mexico of *ahuehuetes*, which owed their prestige to their ancient sacred status in Mesoamerica.[16] It was also at the base of one of these ancient trees that Cortés wept the night he was routed from Tenochtitlán (an episode known in Mexican history as the Noche Triste). The tree seemed to synthesize the history of triumphs and defeats that had visited both parties in the saga of conquest. And because *ahuehuetes* can reach an extraordinary size, they also evoked providential narratives about the bountifulness and unlimited

FIG. 7.4. Chapultepec, *ahuehuete*, and romantic solitude in Pelegrín Clavé, *Paisaje con arbol y río*. From Javier Pérez de Salazar y Solana, *José María Velasco y sus contemporáneos* (1982), image no. 1. The *ahuehuete* and the forest of Chapultepec were staples of the Mexican romantic imagination. As in Germany, the Mexican intelligentsia saw the nation's history deeply connected to both trees and forest. The forest, however, was not the cradle of the Mexican *Volk*. Rather, it triggered tragic recollections about grandiose yet spent precolonial civilizations.

FIG. 7.5. (a) José María Velasco, *Ahuehuete*, in Manuel Ortega Reyes, "El gigante de la flora mexicana o sea el sabino de Santa María de Tule del Estado de Oaxaca" (1882), *La Naturaleza: Periódico científico de la Sociedad Mexicana de Historia Natural*, vol. 6 (1882–84), 110–14; (b) José María Velasco, *Ahuehuete de la Noche Triste* (1885), Museo Nacional de Arte, Mexico City. The gigantic *ahuehuete* pointed to the bountifulness of the land, a sign perhaps of providential election. It also pointed to the historical roots of the nation, sunk deep in the precolonial indigenous past. Finally, the *ahuehuete* was a sign of the brusque historical changes in the nation's past: Cortés was thought to have wept at the foot of one of these giant trees the night he fled Tenochtitlán.

organic potential of the land. Many illustrations and paintings of *ahuehuetes* appeared in Mexico in the nineteenth century, as well as essays elucidating the tree's historical and scientific significance.[17] José María Velasco made the historic symbolism of the tree explicit in his canvas *Ahuehuete de la Noche Triste* (1885) (fig. 7.5).

Two towering figures of nineteenth-century Mexican letters stand out for their contributions to the development of this historically informed type of landscape aesthetics, Manuel Payno (1810–94) and Ignacio Manuel Altamirano (1834–93). Both Payno and Altamirano earned most of their income as journalists, often reporting on their travels across the land. Both in their own ways encouraged the development of a "national" literature in which "Mexican" characters in "Mexican" lands took precedence over stories in undifferentiated settings and locales. Their novels are packed full of national "types," traditions, and landscapes. Not surprisingly, the articles on the places they visited are laced with both historical reportage and passionate aesthetic responses to local nature.[18] Altamirano's *Paisajes y leyendas* (1884), a collection of travel essays, neatly reflect the operation of a mind that appreciates nature mostly through the prism of history.[19]

The very first essay of Altamirano's *Paisajes* captures the logic of this discourse well. After having briefly described the numerous neighborhoods, villages, and other sites along the train ride, Altamirano arrives in the town of Ameca, located on the semi-tropical slopes of the two towering mountains overlooking the central Mexican highlands, Popocateptl and Ixtaccihuatl. Altamirano has come to report on the religious festivities around the cult of an image of Christ in a wooden coffin, El Cristo del Selpulcro, or the Lord of Ameca, miraculously found by the locals a few years after the conquest in a nearby cave, which over time became a chapel carved into the mountain. Given the motif of the image, the mountain had in turn become a Mexican *sacromonte*, visited by huge, rowdy, colorful crowds every Holy Week. Altamirano questioned the historicity of the miracle but praised the festivities as a profoundly beneficial form of communal bonding. Anticipating Émile Durkheim, he saw religion as a way for individuals to bond and worship society itself, and he expatiated on the history of the cult, whose origins he located in the early colonial activities of the Dominicans and Franciscans, who had, in his opinion, forged a new Mexican civilization and identity. Having proved to his own satisfaction that the mid-sixteenth-century hermit Martín Valencia had invented the image and the cult, Altamirano takes on the persona of this Franciscan friar and proceeds to describe the landscape: majestic volcanoes to the north; a sea of undulant valleys and mountains to the south; the striking blue of the atmosphere above; cedars and birds all around. What is remarkable about this passage is that it is not Altamirano himself who provides a lyrical description of the land but a proxy, a sixteenth-century Franciscan. Describing Ameca's landscape through the mouth of a historical character was typical of how nineteenth-century Mexican intellectuals failed to separate aesthetic appreciation of the landscape from concrete historical narratives. Altamirano closes the essay in his own voice, waxing lyrical about the beauty and bountifulness of the land, yet once again ends up linking the description of landscape with a concrete historical figure, namely Sor Juana Inés de la Cruz, the Mexican "tenth Muse": "Ameca ought to be proud of its beautiful sacred hill, and . . . of having sheltered in one of its humble old houses the cradle of that singular celebrity . . . Sor Juana Inés de la Cruz."[20]

Payno and Altamirano strove to make the Mexican public see landscape through the lens of history. How successful they were in doing so can be seen in the pages of one of the most important works of Mexican scholarship of the second half of the nineteenth century, *México a través de los siglos* (1887), edited by the liberal Vicente Riva Palacio (1832–96). The image that introduces each folio volume invariably seeks to make explicit the underlying theme of the work. The frontispiece to volume 1 presents a young woman crowned with olive branches and holding a pen and a book (fig. 7.6), standing for the Mexican republic and its hard-won liberal constitution. Surrounding the beautiful maid, symbols of Mexican identity and history lie scattered about: cactuses, ferns, and palms represent Mexico's various eco-

F I G . 7 . 6 . *(opposite)* Mexico as the mestizo blending of historical periods. Frontispiece to Vicente Riva Palacio et al., *México a través de los siglos* (1887), vol. 1. The most common plant communities of the nation are represented here, thus acknowledging a genre that since Humboldt transformed landscape painting into a form of biodistribution mapping. The history of the nation has its roots firmly in this landscape: the three most important historical epochs of the nation are acknowledged. The new nation, a woman clad in white and holding a pen and a book, a constitution, is flanked both by the Indian and the Spanish pasts. A new mestizo type stands next to the Spanish conquistador. The founding fathers of the liberal nation are many. They are all liminal figures, turning points in the tripartite structure of the nation's past: Moctezuma, the last of the Aztec kings; Hernán Cortés, the conquistador who brought the Aztec empire down; Miguel Hidalgo, the priest who brought the Spanish empire down; and Benito Juárez, who ushered in the new liberal nation.

EDITORES. BALLESCÁ, ESPASA Y COMP.ñia

logical habitats; precolonial objects signify the grandeur of the Amerindian past. A historical narrative unfolds as well. Behind the maiden hangs the seal of the Mexican nation, a purported Aztec depiction of the prophetic and providential origins of the city of Tenochtitlán, in which an eagle perched on a nopal devours a serpent. A series of portraits of Moctezuma, Cortés, Hidalgo, and Juárez suggests the periodization used to organize Riva Palacio's mammoth editorial effort: ancient history, colonial history, wars of independence, civil wars, and liberal reforms. The portraits also reveal the evolutionary narrative of historical continuity underlying the work as a whole. A narrative of radical change, however, unfolds parallel to (and right beneath) the evolutionary tale of progress. Two brown-skinned Amerindians and a white-bearded Spanish conquistador stand next to a Mexican mestizo, whose deportment and costume differ as radically from the Amerindians' as from the Spaniard's. The narrative of Mexico as a mestizo civilization is reinforced by the posture of the white maiden, for the nation's visage is turned toward the Amerindians. *México a través de los siglos* argues that the nation has evolved from an ancient, grandiose yet barbarous society into a liberal polity, in which constitutional principles and the law rule at last. The book also suggests that no past historical period can be dismissed or belittled, for a new mestizo Mexican civilization emerged out of the encounter of Amerindians and Spanish conquistadors. Mexico should stop running away from both its Amerindian and Spanish pasts.

Landscapes in the frontispieces to each volume help bolster this argument. The most striking is perhaps the introductory image to the volume on history of the colonial era. *México a través de los siglos* sets out to recover the colonial past from the condescension of nineteenth-century liberal intellectuals, who took it to be a period characterized by medieval obscurantism, inquisitorial repression, and cruel Spanish domination, an era to be forgotten rather than celebrated. The frontispiece of volume 2, however, seeks to make obvious the role of the Church in the crafting of the new mestizo nation (fig. 7.7). In the background stand the two volcanoes that overlook the valley of Mexico, Popocateptl (smoking mountain) and Ixtaccihuatl (sleeping woman). The choice of image is deliberate, for the volcanoes had long been linked in the collective imaginary with a purported Aztec myth of *mestizaje*.

In an illuminating book, Keith H. Basso has explored how the Apaches used place-names to store and articulate complex narratives of the past.[21] Scholars in Mexico were very well aware of these indigenous traditions of using landscapes as shorthand for historical narratives. Elsewhere I have described the case of eighteenth-century Mexican baroque antiquarians

FIG. 7.7. Cultural *mestizaje* and landscape. Frontispiece to Vicente Riva
Palacio et al., *México a través de los siglos* (1887), vol. 2. The colonial period
is represented here as a period of cultural blending and racial *mestizaje*. A
conquistador and an Indian woman bring a new mestizo child to be baptized
by a friar, while two other natives and a mestizo witness the event. Missing
in this scene of cultural and racial *mestizaje* is any mention of blacks, who
arrived in Mexico as slaves by the hundreds of thousands. The landscape
itself (the volcanoes of the central valley of Mexico) helps reinforce the
message of *mestizaje*.

FIG. 7.8. Saturnino Herrán, *La leyenda de los volcanes* (1910). Pinacoteca del Ateneo Fuente de Saltillo, Saltillo, Coahuila. Mexicans built on a well-established indigenous tradition of storing complex historical memories into place-names. The names of the volcanoes Popocatepetl and Ixtaccihuatl were supposedly those of two Aztec lovers. "Sleeping White Woman" (Ixtaccihuatl) unintentionally commits adultery when she thinks that her departed husband, the Aztec warrior "Smoking Mountain" (Popocatepetl), has died. Both Ixtaccihuatl and Popocatepetl commit suicide when the warrior comes home and the adultery becomes clear. From the corpses of the lovers arose the two volcanoes.

who made extensive use of place-names as alternative sources from which to create a new historiography of the Western Hemisphere.[22] In the nineteenth century, Mexican intellectuals used the names of Popocateptl and Ixtaccihuatl to create a narrative of primeval Mexican *mestizaje*: Upon returning from a failed military campaign of many months, an Aztec soldier finds his wife, who is as beautiful as a white lily, remarried to a treacherous Tlaxcalan warrior, who has tricked the woman into believing that the Aztec had died in battle. The Aztec routs the Tlaxcalan, but in the meantime, the dishonored wife commits suicide. As the grieving Aztec warrior embraces his wife's white corpse, the skies darken and a geological catastrophe ensues. The next day, the valley awakens with two new mountains: Ixtaccihuatl (the "sleeping" white woman), and Popocatepetl (the Aztec warrior). The two tragic lovers have been converted into features of the landscape.[23] The image of the volcanoes in the frontispiece to Riva Palacio's second volume was thus not casual. It deliberately sought to reinforce the thesis of *mestizaje*

FIG. 7.9. Luis Coto, *Hidalgo en el Monte de las cruces* (1879). Museo de Arte, Instituto Mexiquense de Cultura, Toluca.

that the book as a whole was trying to convey. The trope of *mestizaje* in the volcanic landscape introduced by Riva Palacio was given expression in 1910 by Saturnino Herrán's modernist painting *La leyenda de los volcanes* (The Legend of the Volcanoes) (fig. 7.8).

The frontispieces to the other volumes, which trace the history of Mexico from the wars of independence to national decline to a jubilant, prosperous liberal present, also convey their messages by means of landscape symbolism. Three images tell a tale of national resurrection, fall, and regeneration. In the frontispiece to the third volume (on the war of independence), the Amerindian spirit of the nation rises free, Phoenix-like, from the ashes of Mexico's battlefields (fig. 7.10). The background to this image is a scene of the combat waged by the patriot armies of Manuel Hidalgo y Costilla against the Spanish royalist armies in 1810 at the Monte de las cruces, a forested mountain north of Mexico City. The obelisk gives the locale away, for in this very volume, in the section devoted to this battle, there is a landscape of the site with an identical obelisk, built to commemorate Hidalgo's triumph (fig. 7.11). These two images of the same landscape should be contrasted with that exhibited a few years earlier in the Academy of San Carlos's 1879 salon by Luis Coto (1830–1889) (fig. 7.9), whose rendition of it was coldly received.

FIG. 7.10. Mexico rises from the ashes of the wars of independence. Frontis-piece to Vicente Riva Palacio et al., *México a través de los siglos* (1887), vol. 3.

FIG. 7.11. The *Monte de las cruces* in Vicente Riva Palacio et al., *México a través de los siglos* (1887), vol. 3.

Of all people, Ignacio Altamirano, who had found it so difficult to sever history from the aesthetic description of landscape, took Coto to task for having invented a new hybrid genre: neither a landscape nor a historical scene. Coto could not make up his mind between these two genres, Altamirano charged, and the end result was a painting of a badly framed generic forest swamping an ill-focused historical scene.[24] The two images of the Monte de las cruces in *México a través de los siglos* seek to heed Altamirano's critical advice (figs. 7.10 and 7.11). One is clearly a landscape, the other a historical scene; both however, exist to edify and educate the nation.

The frontispiece to volume 4 (on the period of anarchy that followed independence) announces why the nation collapsed (fig. 7.12). The image is of a maiden clad in red and wrapped in an ominous black shawl, clutching a dagger and a torch: the nation has turned against itself. A landscape of a city in flames and of stumps and manacles, circling vultures, and tombs of the nation's founding fathers speak of the desolation brought about by economic decline, internecine wars, and foreign occupations. Although the images of a rainbow and of a tree struck by lightning suggest the hope of a future

FIG. 7.12. Desolate nineteenth-century Mexico. Frontispiece to Vicente Riva Palacio et al., *México a través de los siglos* (1887), vol. 4. This image captures the crisis Mexico went through in the course of the nineteenth century: territorial dismemberment, economic collapse, civil wars. Mexico is represented by a dark, ominous figure clutching a stabbing knife and a torch to set Mexico on fire. Manacles, stumps, tombs, lightning bolts, and vultures point to the depressing state of post-independence Mexico. The rainbow in the horizon, however, announces hope.

regeneration, the landscape is bleak and barren and symbolizes Mexico's fall in the nineteenth century.

The frontispiece to the final volume, however, presents a radically altered panorama: a triumphant liberal nation, clad in the colors of the Mexican flag and holding in her hands the promises of the new republic—peace (an olive branch) and the rule of law (the tablets of the liberal constitution of 1857)—flies through a pristine, blue sky among cirrus and cumulus clouds (fig. 7.13). This image is also a landscape, albeit of clouds, part of a well-established nineteenth-century genre depicting cloud formations.[25]

Landscapes and history are intimately connected in the pages of *México a través de los siglos*. In fact, landscapes not only appear in the frontispieces to each volume but also surface periodically as illustrations to many of the book's essays. Thus, for example, an account of the long-enduring Toltecs is followed by a sunset amid the ruins of Teotihuacan; the story of the fifteenth-century Acolhua ruler Nezahualcoyotl is complemented by one

FIG. 7.13. Liberal constitution and heavenly landscape. Frontispiece to Vicente Riva Palacio et al., *México a través de los siglos* (1887), vol. 5.

of Velasco's views of the site where the legendary monarch had ordered his baths built; the narrative of the conquest is followed by an engraving of the *ahuehuete* of the Noche Triste; the tale of the wars for independence waged by Hidalgo is complemented by a scene of the Monte de las cruces; a biography of the priest José María Morelos, a hero of the wars of independence, contains a panoramic view of the valley of San Cristobal Ecatepec, where Morelos was dispatched by a royalist firing squad; finally, a description of the U.S.-Mexican War includes an image of the lands of the monastery of Churubusco, where a battle was fought.

The Religious and Scientific Worlds of José María Velasco

José María Velasco inherited this tradition of landscape aesthetics. Despite the many volumes published about him, Velasco and his work still remain poorly understood. A provincial, born in a small town to the north of Mexico

City into a family of small merchants, Velasco arrived in the capital relatively young. He gave early signs of being a skillful draftsman and enrolled in the Academy of Art at the age of eighteen (1858). Had he registered a few years earlier, he might have become an exceptionally talented portraitist or a painter of historical and biblical scenes, but as it was, he entered an academy undergoing momentous changes, including the hiring in 1855 of an Italian landscape painter, Eugenio Landesio (1810–79). In addition to a love of neoclassical pastoral scenes and the Roman Campagna, Landesio brought to Mexico an aesthetic sensibility open to new experiences. He soon found himself with commissions to paint scenes of far-flung haciendas for wealthy patrons. A cadre of young painters flocked around the Italian, including Velasco, Luis Coto, and Gregorio Dumaine, two other influential nineteenth-century Mexican landscape painters. Landesio took the students to the countryside to sketch, sent them on expeditions to geologically interesting sites, and used them as draftsmen and engravers to illustrate two of his books.[26]

In time, Velasco developed into one of the leading natural historians of Mexico. Like his counterparts to the north, the members of the Hudson River School of painting, he was thoroughly steeped in the science of his age. His paintings of trees, plants, land formations, and clouds reveal an exquisite knowledge of botany, geology, and meteorology. Like Frederick Edwin Church (1826–1900), Velasco approached nature as an alternative to the Bible and set out to find sacred revelation in the landscape.[27] Unlike Church, who mostly pursued a contemplative form of natural theology, Velasco was a hands-on natural historian. Along with his brother Idelfonso, he kept a botanical garden, where he conducted experiments that culminated in botanical publications.[28] His commitment to science led him to rise through the ranks of the learned, and he in turn became secretary (1880–81), vice-president, and president (1881–82) of the Mexican Society of Natural History, whose official organ, the journal *Naturaleza*, he helped launch, contributing articles and illustrations.

Like any other "Victorian" bourgeois, Velasco wrestled with the erosion of providential narratives by nineteenth-century evolutionary ideas. Although steeped in the science of his age, particularly geology and natural history, Velasco was a deeply conservative and religious man who found the sacred in nature, and he rejected most of the insights of the Darwinian revolution. He believed he had found evidence supporting his anti-evolutionary views in a curious Mexican salamander, the axolotl, or *ajolote* (fig. 7.14). Velasco targeted the theories of the German biologist August Weismann (1834–1914).

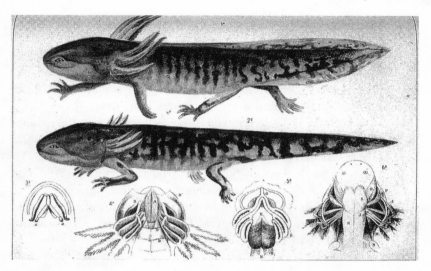

FIG. 7.14. Cuverian study of the axolotl, or *ajolote*, in José María Velasco, "Descripción, metamórfosis y costumbres de una especie nueva del género Siredón encontrada en el Lago de Santa Isabel, cerca del Villa de Guadalupe Hidalgo, Valle de México" (1878). Velasco was a sophisticated naturalist familiar with Cuverian comparative anatomy and with recent developments in Darwinian evolution. His study of the axolotl challenged Darwinian theories of evolution as articulated by the German August Weismann. Politically, Velasco was a conservative. His Catholic religiosity made him hostile to Darwinism and evolution. As a naturalist, Velasco used his paintings to work out detailed landscape studies of rock formations, plant communities, and cloud forms.

Drawing upon several studies of the toadlike axolotls of central Mexico by French naturalists like Georges Cuvier (1769–1832) and Auguste Duméril (1812–70), Weismann had concluded that these creatures showed how species evolved. Axolotls are, in fact, a type of larval salamander, exhibiting neoteny, failing to undergo metamorphosis to the adult stage and retaining gills and fins. Yet if forced to survive outside the water, their bodies, eyes, and tails change, and they develop full-fledged lungs, while their gills atrophy. Such a transformation appeared to support Weismann's thesis of a "phylogenic," progressive direction to the seemingly random process of natural selection, although when axolotls became salamanders, they seemed to regress to a youthful stage, because they became sterile, which suggested degeneration rather than progress. Weismann argued, however, that the aquatic axolotls must originally have descended from terrestrial salamanders, and that their transformation into sterile salamanders was simply a reversion to their earlier phylogenic form. Velasco sought to discredit the neo-Darwinian ideas of the German biologist. Paradoxically, in a brilliant earlier study in the best tradition of Cuvier's morphological anatomy, Velasco had argued that the axolotls of Mexico did indeed mutate randomly into salamanders.[29] In his rejoinder to Weismann, Velasco sought, however, to prove that the random mutations were not the result of changes in environmental conditions (water level in the lakes), and that the resulting salamanders were not sterile. The changes were innate to the species of axolotls, he argued, and could not be taken as proof of directionality in any alleged evolutionary process (either of progress or degeneration). Velasco insisted that Weismann's errors stemmed from the type of experimental biology the German practiced, for Weismann had raised his axolotls in captivity. Velasco depicted himself as a privileged observer who had studied axolotls in situ, that is, in their undisturbed natural settings.[30]

The Central Valley as Metonym of the Geography of the Nation

This worldly biologist strangely comes across as a landscape painter with a somewhat limited geographical repertoire, for he painted only a handful of landscapes outside the central highlands, from the central valley of Mexico to the neighboring valleys of Tlaxcala, Puebla, and Temascalcingo, his hometown. Velasco almost reluctantly traveled to the states of Oaxaca, Veracruz, and Querétaro to paint. His fame took him to world's fairs in Philadelphia, Paris, and Chicago, but all he left from these trips are homesick letters addressed to his wife and a couple of seascapes from the port of La Havana.[31] Compared to Frederic Church, Velasco comes across as a provincial. Church's cosmopolitan repertoire ranges from icebergs off the coast of Newfoundland to the wilderness of New England to the tropics of the Caribbean to the highlands of Ecuador and to arid landscapes of the Middle East. Yet these differences in geographical range simply betray differences in ways of being imperial.

Church's cosmopolitan stare speaks to a view of the nation as an ever-expanding "frontier." Clearly in tune with the ideology of "Manifest Destiny," Church turned his gaze to the Caribbean and South America in the same spirit with which Albert Bierstadt (1830–1902), that other giant of mid-nineteenth-century U.S. landscape painting, approached the West, namely, as a "wilderness" to be both tamed and revered. The painters of the Hudson River School used the trope of the "wilderness" not only to naturalize the imperial vocation of the adolescent U.S. nation but also to give it a history. Canyons and forests in the wilderness acted as substitute historical monuments for a youthful people who did not have a Parthenon or medieval cathedrals to brag about. Church, nevertheless, reached across the Atlantic to anchor the nation's historical identity in other places, namely, in England, Greece, and the biblical lands. The generous geographical range in Church's landscapes, therefore, resulted from the problematic construction of a hegemonic national identity that in nineteenth-century United States was imperial, classically rooted, Judeo-Christian, and Anglo, all at the same time.[32]

Velasco's landscapes, on the other hand, reveal a view of the nation as urban and rooted in the central Mesoamerican highlands at the expense of everything else. Strangely absent from nineteenth-century Mexican landscape traditions, with only a handful of exceptions, is a rural view of the land, of frontiers, cattle, and cowboys.[33] This is all the more strange inasmuch as this period witnessed the massive settlement and development of Mexico's long loosely held frontier. U.S. audiences, so ready to see Mexico as a land of impoverished peasants and well-mannered aristocrats ensconced in their feudal rural haciendas, would most likely be surprised to discover that most landscape paintings of rural estates in nineteenth-century Mexico tended to concentrate on the buildings, not the surroundings.[34] The notion of the wilderness, by and large, is absent from the visual repertoire of Mexican painters; illustrations of valleys and highlands consistently have a town or a city as their focus. It would, moreover, be an error to attribute this urban-centered sensibility to the imposition of Mediterranean urban traditions on the Amerindian poor by arriviste, neocolonial elites. Ethnohistorians have repeatedly noted that this urban focus was also central to the corporate and ethnic identities of the disenfranchised "rural" Amerindian population.[35]

Whose Imperial Eyes?

Mary Louise Pratt has argued that Alexander von Humboldt's aesthetics had a lasting influence in nineteenth-century Latin America. According to Pratt, Humboldt's representations of a lush, exuberant tropical America captivated Latin American elites desperate for European validation.[36] Pratt's thesis marvelously fits our expectations of the nineteenth-century Latin American comprador bourgeoisie: hopelessly derivative. But Pratt is wrong. At least in Mexico, intellectuals forging a national landscape furiously fought *against*

FIG. 7.15. Mexico as empty space. Johann Moritz Rugendas, *Plateau of Puebla*, in Carl Christian Sartorius, *Mexico: Landscapes and Popular Sketches* (1859). Like his mentor, Alexander von Humboldt, Rugendas favored typical landscapes (vegetation, rock formations, etc.), and his emphasis on biodistribution led him to depict the central valley of Mexico as empty.

the type of landscape aesthetics first introduced by Humboldt. Humboldt bequeathed to his followers a view of tropical landscapes as spaces to study biodistribution, full of diverse plant and animal populations, but empty of humans. It was the Hudson River School of painting that actually embraced Humboldt's aesthetics of the tropics, not Latin American painters.[37]

The urban-centrism of nineteenth-century Mexican landscape painting was exacerbated by the need to respond to the condescending views held by foreigners made widely available in Europe and the United States in travel accounts. It could be argued that the tendency to merge cityscapes with landscapes originated as a response to foreign views of Mexico as a "wilderness." Take, for example, the text published both in English and German by Carl Christian Sartorius, *Mexico: Landscapes and Popular Sketches* (1859), with illustrations by the German traveler-artist Johann Moritz Rugendas (1802–58), which, among other things, seeks to identify landscape types in Mexico, ranging from alpine to tropical. Rugendas and Sartorius went to the New World to fulfill Humboldt's scientific and aesthetic programs. The

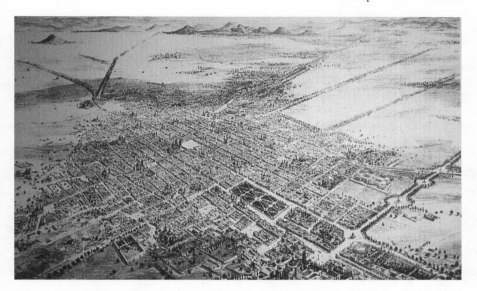

FIG. 7.16. Mexico as cityscape in *México y sus alrededores* (1855–56). The intelligentsia of the capital often took the central valley of Mexico to be the nation. This was clearly an imperial attitude that marginalized all other regions. This imperial attitude was challenged by provincial liberals in particular. There was no discussion, however, when it came to representing the nation through landscapes; both provincial liberals and imperial conservatives from the capital assumed landscapes to be cityscapes. Mexicans fought back furiously against Humboldtian representations of Mexico as wilderness, an empty space available for the taking.

end result of this exercise was a book full of illustrations depicting Mexico as an "empty" rural wilderness (fig. 7.15). These views contrast dramatically with those offered in *Mexico y sus alrededores* (1856–57), where the nation is primarily urban, the capital stands for the whole, and aerial views of Mexico City merge the cityscape with the landscape of the central valley (fig. 7.16).

In his merging of landscapes with cityscapes, Velasco was typical. Less typical, however, was his tendency to paint mostly views of the capital. Here, Velasco clearly cast his lot with the program of Mexican conservatives. The history of nineteenth-century Mexico could simplistically be represented as the clash between two agendas, namely, that of the provinces, particularly the semicrescent around the capital, from San Luis Potosí to Guadalajara to Oaxaca, which fought for the dismantling of the colonial ancien régime and the expansion of the franchise to all males, and that of the capital and its port, Veracruz, which advocated a hierarchical, aristocratic view of the polity, longing for a monarchical solution to the secular crisis of the nineteenth century.[38] Critics of Velasco and his circle made the political agenda underlying Velasco's limited geographical repertoire explicit. Ignacio Altamirano, himself a provincial "Indian" from the state of Guerrero and a leading liberal intellectual, for example, harshly criticized Velasco for the narrow range of the latter's geographical imagination and for Velasco's tendency to confuse the whole of Mexico with the capital.[39] The Mexican Society of Geography and Statistics, long a front for the liberal intelligentsia, on the other hand, denounced Velasco's Society of Natural History for espousing the intellectual values the ancien régime.[40]

It should be clear by now that Velasco harbored deeply conservative views of society, religion, and nature, which colored the way he represented the nation in his paintings.[41] But for all his conservative leanings, Velasco's landscapes are not substantially different from those offered by more liberal figures like Altamirano or Riva Palacio. Like Altamirano and Riva Palacio, Velasco used nature to craft a mestizo image of the nation, a view of the land that recuperated both the indigenous and the colonial past through its landmarks. In the following pages, I offer a reading of Velasco's various views of the central valley of Mexico and argue that these paintings should be read as a historical narrative about Mexico's new mestizo identity.

The Central Valley as Metonym of the History of the Nation

As the Mexican art historian Fausto Ramírez has indicated, Velasco carefully composed his landscapes for at least twenty years of his career (1868–89) to convey some sort of historical narrative. In the 1870s, Velasco created his first compositions of the central valley of Mexico, full of references to its history. Due to his obvious historical interests, the National Museum, an establishment founded in 1831 to study the great Amerindian civilizations of the precolonial past, appointed him its official photographer and illustrator in 1877. For three decades, Velasco made copies of precolonial codices, images of the museum's growing archeological holdings, and illustrations for *Anales*, the museum's periodical. This intense activity, in turn, reinforced his original interest in evoking history in landscapes. Ramírez has astutely pointed out that most of Velasco's 1880s landscapes have historical themes, ranging from his obvious compositions of the pyramids of Teotihuacan and the *ahuehuete* of the Noche Triste to the more subtle panoramas of the state of Oaxaca. Each and every one of his Oaxacan paintings, for example, was organized around

FIG. 7.17. Layers of history in landscape, I. José María Velasco, *View of the Valley of Mexico from the Hill of Atzacoalco* (1873). Private collection. Velasco saw landscape as a palimpsest of historical memories. Like most Mexican intellectuals, he understood Mexican history to be divided in three clear-cut historical ages: precolonial, colonial, and republican. Velasco's landscapes used the layers of the canvas to transmit this message. The foreground in this case refers to the colonial roots of the nation: the cult of Our Lady of Guadalupe. This is a northern view of the central valley. In the northern foreground, firmly associated with the indigenous roots of the nation, the source of many Indian migrations into the valley in the precolonial period, a group of natives gather to venerate an image of the Our Lady of Guadalupe. The middle ground is occupied by shrines to the Virgin and her miracle on the hill of Tepeyac. The background is the valley itself, crisscrossed by railroads, where the capital of the nations is nestled. The deep background is occupied by the volcanoes, pointing to the mestizo roots of the nation.

a historical theme, including the highlands of Guelateo (to commemorate Benito Juárez's birthplace), views of La Carbonera (to celebrate the place where Porfirio Díaz defeated the French in 1866), and panoramas of Mitla (to evoke the great precolonial Zapotec civilization).[42]

Velasco is known mostly for his panoramic views of the valley where Mexico's capital is nestled amid lakes and surrounded by ice-capped volcanoes. Velasco painted the valley from every conceivable direction (figs.

F I G . 7.18. Layers of history in landscape, II. José María Velasco, *View of the Valley of Mexico from the Hill of Santa Isabel* (1877). Museo Nacional de Arte, Mexico City. The same historical narrative as in figure 7.17 that privileges a northern perspective of the valley is presented here. Again, each segment of the canvas represents a historical period. But in this case, the tripartite historical narrative of the Mexican past is clearly revealed through each segment. The cactus and the eagle in the foreground point to the precolonial Aztec roots of the nation. The middle ground again points to Tepeyac and the shrine of Our Lady of Guadalupe, as metonym of the colonial period. The vibrant capital and the valley crisscrossed by railroads point to the republican modernity.

7.17–19). He first drew it from the mountains to the north, offering breath-taking panoramas. Over the years, however, he lightened his palette and painted the valley from ground level, presenting more intimate, less sweep-ing perspectives from the east, west, and south.

This switch corresponded to his partial abandonment of historically charged landscape symbolism as he unsuccessfully sought to maintain his prestige with new generations who were increasingly attracted to more fashionable, modernist aesthetic sensibilities. Nevertheless, Velasco is best remembered for his almost aerial views of the valley from the northern

FIG. 7.19: Volcanoes, catastrophes, and *mestizaje*. José María Velasco, *Popocateptl e Ixtaccihuatl desde el lago de Chalco* (1882). Private collection. Here Velasco has abandoned the northern perspective and the historical metaphor through the play of segments of the canvas. The arrival of modernism and the avant garde forced Velasco to modify both his palette and his repertoire. Modernism deliberately abandoned the use of landscape to offer historical narratives of the nation. Velasco, however, continued to use the volcanoes of the valley as a metonym for the mestizo roots of the nation.

mountain ranges. Through the use of geographical landmarks and the manipulation of the various planes in the painting, he sought to convey subtle historical narratives about the nation, which have yet to be fully understood.

Velasco was not the first painter to have offered sweeping, panoramic views of the central valley of Mexico. Foreigners like Rugendas, Jean-Baptiste Louis Gros (1793–1870), Daniel Thomas Egerton (1800–1842), and Conrad Wise Chapman (1842–1910), and Mexicans like Pedro Calvo (fl. 1825), Urbano López (fl. 1840s), Casimiro Castro, Eugenio Landesio (1810–79), and Salvador Murillo (fl. 1860s) had already painted the valley. In addition to his undeniable technical mastery, what Velasco seems to have brought to the genre was a desire to evoke the valley's history. In this, to be sure, he was not alone. Most nineteenth-century poetic descriptions of the central highlands tend to be aerial, sweeping panoramas that quickly mutate into historical narratives. The poems "En el Teocalli de Cholula" (1820) by José María Heredia and "A la vista del Valle de México" (1860) by Ramón I. Alcaraz are cases in point. After waxing lyrical about the beauty and bountifulness of the valley, Heredia jumps to a discussion of the geological history of the valley. Geological landmarks like Popocateptl, he argues, have witnessed the waves and tides of civilizations unperturbed. Based on contemporary vulcanist geological histories of the Earth, Heredia speculates that even the mountains will one day implode and disappear: "Everything passes away according to universal laws. Even this world that we inhabit, so bright and beautiful, is the pale deformed corpse of a previous one."[43] From deep time, Heredia then switches to a discussion of history on the human scale, using an Aztec pyramid in the valley as a pretext. The pyramid helps him evoke the pageantry and foolishness surrounding human sacrifice. The long-enduring pyramid, Heredia concludes, should remind Mexicans of the acts of demented furor that human nature is always capable of committing. Alcaraz's poem is much longer than Heredia's, but it basically follows the same structure: an aerial, sweeping panoramic description of the valley and the city is followed by a vulcanist geological history and by a painstaking description of the various civilizations that the land has seen rise and fall, from the Toltecs to the Chichimecs to the Aztecs to the Spanish conquest to the wars of independence to the postcolonial fratricidal conflicts.[44]

In the tradition of Heredia and Alcaraz, Velasco first sought to offer a geological narrative of the valley's history. The ice-capped volcanoes of the central valley are a silent background to all his compositions. Like the painters of the Hudson River School, Velasco kept abreast with the most recent geological debates and left countless studies of rock formations. But unlike the painters of the Hudson River School who left numerous clues as to their preferred geological theories in their paintings (erratic boulders as evidence of the Flood in Thomas Cole's case; compositions on the geological forces of erosion and sedimentation supporting the uniformitarian views of Charles Lyell in Church's case; and seascapes elucidating the role of Louis Agassiz's glaciations for rock formations in William S. Haseltine's case), Velasco left no clues as to his favorite geological views.[45] For all his criticism of evolutionary ideas, Velasco never denied the revolution in the perception of historical time introduced by nineteenth-century geology. He in fact painted murals illustrating the Earth's various geological ages for the Museum of Geology that Porfirio Díaz inaugurated in 1906 to host the Tenth International Geological Congress. It is therefore plausible to argue that Velasco saw the geological history of the valley through the eyes of a mind informed by Cuvier's theories: fossils were the result of successive catastrophic extinctions, which were followed by new, divinely inspired creations. A close reading of Heredia's and Alcaraz's and countless other descriptions of the valley indicate that the nineteenth-century Mexican intelligentsia favored catastrophic geological narratives, particularly vulcanist ones.[46] Clearly, Mexican intellectuals preferred catastrophist geological narratives to uniformitarian ones largely because their reading of the history of the Earth was colored by the tumultuous and tragic events they experienced in the nineteenth century. The Mexican literate elites understood geological change largely as the result of volcanic activity: sudden, massive earthly convulsions. It is not surprising, therefore, that intellectuals like Velasco lavished so much attention on Popocateptl and Ixtaccihuatl. Moreover, as good romantics, they perceived the mountains through the aesthetics of the sublime, experiencing the beautiful in awe and terror. The haunting presence of Popocateptl and Ixtaccihuatl in Velasco's landscapes perhaps also suggests symbolic readings: the volcanoes as emblems of *mestizaje*.

Like Riva Palacio, Velasco subscribed to a mestizo narrative of the history of Mexico. Velasco's *Valley of Mexico from the Hill of Atzacoalco* (1873) subtly uses the landscape to convey these views (see fig. 7.17). In the foreground, Amerindians gather around the image of Our Lady of Guadalupe amid giant boulders and nopales and other typical flora of the Mexican highlands. The hill of Tepeyac to the north of the capital, with its shrines and cathedrals, where Our Lady of Guadalupe allegedly first appeared, stands in the middle ground. The valley, crisscrossed by railroads, the city, and the southern mountain ranges of Ajusco constitute the background. In addition to blending indigenous, colonial, and modern symbols that represent different periods of Mexican history, Velasco turns the landscape into a historical narrative simply by his choice of a northern perspective. The north had long been associated in central Mexico with change and renewal, and with a series of barbarian migrations that had caused civilizations to fall. Alcaraz's poem to the valley of Mexico is typical of this genre, for he describes in great detail the many cycles of civilizations witnessed in the valley owing to the various northern barbarian invasions. This narrative of the rise and fall of civilizations was appropriately adjusted in the nineteenth century to account for the dismal "downward" trajectory of Mexico after the wars of independence. Mid-nineteenth-century poets like Manuel Carpio deployed the trope to explain the loss of territory to the United States. Carpio casts Mexico City as the Rome of the Western Hemisphere, threatened by "Norman," northern barbarians. The internecine struggles and the effeminization of local elites rendered the Mexican imperial city helpless and led to the U.S. barbarian invasion.[47] Like Carpio and Alcaraz, Velasco privileged Mexico City as the point where all historical narratives converged, as the Rome of the Western Hemisphere. But for Velasco, the Mexican Rome had a colonial mestizo past, the result of the indigenous embrace of the rituals and beliefs of the European Catholic Reformation embodied in the image of Our Lady of Guadalupe.

The role of Our Lady of Guadalupe in the formation of a distinct mestizo Mexican identity is well documented.[48] Originally conceived as a colonial Creole patriotic cult, the veneration of the image slowly developed into a pan-regional indigenous devotion, creating a genuinely national practice that celebrated a providential Mexican destiny. The providential religious narrative is reinforced in Velasco's painting by the contrast between the brown, barren mountains of Atzacoalco (foreground) and Tepeyac (middle ground) and the green, bountiful pastures of the valley below. Here Velasco seems to be drawing upon well-established aesthetic conventions. From its inception, the devotional literature of the colonial period describing the miracle created a poetic genre that elucidated the meaning of the miraculous appearance through the metaphor of environmental transformation: while Tepeyac was barren, the valley below was bountiful. Our Lady of Guadalupe had chosen the barren hill of Tepeyac in 1531 to signify her promise of the rich spiritual and material harvests to come.[49]

Velasco's *Valley of Mexico from the Hill of Santa Isabel* (1877) deepened the historical narrative he first introduced in 1873 (see fig. 7.18). In the 1877 version, the valley is again seen from the north. This time, however, the hill of Atzacoalco has receded into the middle ground along with the hill of Tepeyac, and Popocateptl and Ixtaccihuatl stand in the background. In the foreground, a nopal and an eagle on the slopes of the hill of Santa Isabel make transparent the connection with the Aztec past through the myth of the providential foundation of the city of Mexico. This painting represents Velasco's most complete and sophisticated statement about Mexico's mestizo identity. The foreground stands for the continuous presence of the Mexican past; the middle ground speaks to the colonial, Catholic roots of the nation; the deep background manipulates narratives of *mestizaje* embodied in the Aztec legend of the volcanoes; finally, the city nestled amid the valley's lakes speaks to the urban and imperial vocation of the nation as interpreted in nineteenth-century conservative circles.

The Historical Landscape and a New Discourse of *Mestizaje*

The discourse of *mestizaje* hammered out by Velasco and his peers had little to do with racial miscegenation. Mauricio Tenorio-Trillo and Stacie G. Widdifield have contributed significantly to an understanding of how discourses of *mestizaje* were deployed and manipulated by the nineteenth-century Mexican elite. According to Tenorio-Trillo, Mexican intellectuals hesitated to accept the hegemonic North Atlantic discourses in which racial mixing was thought to cause degeneration. Although local intellectuals did think that contemporary indigenous peoples and poor mestizos were degenerate, they assumed that this was a result of their lack of education and poor living conditions, not of any innate racial inferiority. Social and cultural factors, not racial mixing, were the cause of Mexico's plight. This skeptical view of the dangers of racial miscegenation, Tenorio-Trillo suggests, allowed them to embrace Mexico's glorious Amerindian past wholeheartedly. Without embarrassment, they presented an image at World's Fairs of contemporary Mexico as heir to the spent grandeur of the Aztecs and Toltecs.[50] Widdifield, on the other hand, has shown that Mexican artists and intellectuals inverted the traditional view of *mestizaje*, namely, one in which the dominant white male takes over a female racial other as sexual partner. The literate elite embraced a view of the nation made up of brown "Indian" males seduced by the attractions of a female white European civilization.[51]

But for all their insights into *mestizaje* in nineteenth-century Mexico, Tenorio-Trillo and Widdifield do not adequately capture the impulse behind the historical and landscape narratives that I have so far described. The landscapes of Riva Palacio and Velasco, to cite just two examples, have only partially to do with racial miscegenation. The authors and painters I have studied sought to offer a new master narrative for the nation, one that did not shy away from the whole of the Mexican past. In light of the secular tendency in nineteenth-century Mexico to dismiss the precolonial and colonial pasts as a cultural burden, the new historical narratives of Riva Palacio and Velasco sought to do the opposite, namely, to incorporate purportedly radically different historical stages into a single evolutionary narrative of progress. The nation, these intellectuals suggested, needed to stop running away from its indigenous and European roots. In their eyes, the nation was neither "Indian" nor "Spanish"; rather it belonged to a new historical stage. According to this discourse, *mestizaje* was a new (third) historical phase in an evolutionary teleology of progress. *Mestizaje* was a synonym for "modernity": a willingness to bring (and critically embrace) new forms of political organization, technologies, and economies into the fold of a community whose contours, as much as its landscapes, had been shaped by distinct historical stages and events.

In the case of Velasco, this curious view of "historiographical *mestizaje*" was reflected pictorially in layers and was never articulated in writing. It appears that for Velasco, each segment of the canvas was meant to represent a different historical period. Since his views of the central valley of Mexico were northern, aerial ones, the foreground usually stood for the precolonial period (the source of Chichimec, Toltec, and Mexica migrations). The colonial period, most often, was represented by the middle ground, the space occupied by the hill of Tepeyac and the buildings and institutions created there by the Catholic Church around the cult of Our Lady of Guadalupe. The background was the central valley itself, standing for a longed-for modernity, occupied by a great metropolis, crisscrossed by railroads. The whole, however, was always set within the deep background of the volcanoes of the Sierra Madre, which in the nineteenth-century Mexican imagination came to stand for Mexico's racial and cultural hybridities alike.

The Modernist Unraveling of the Historical in the Landscape

By the 1880s, Velasco was slowly moving away from his sweeping northern panoramas. As previously noted, his historically charged landscapes began to give way to more intimate, partial views of the valley. (However, he continued to churn out copies of his 1870s panoramas to meet overseas demand.) This transition seems to have been part of a larger reorientation in aesthetic sensibilities among the Mexican intelligentsia. In the case of Velasco, he changed as he struggled to keep his influence and prestige. But for all his efforts to adjust, by the early twentieth century, he was forgotten and the new discourse of modernism became hegemonic. The impressionism and neo-impressionism of painters like Joaquín Clausell (1866–1935) soon displaced the old master.[52] This fin de siècle transformation spelled the temporary abandonment of the genre of historically informed landscapes. It took the revolutionary struggles of the 1910s to revive and revamp the genre, this time clad in the modernist idioms of Gerardo Murillo (aka Dr. Atl) (1875–1964). In this section, I briefly sketch the rise of this new ahistorical sensibility by looking at the writings of Joaquín Arcadio Pagaza (1839–1918) and Manuel José Othón (1858–1906).

Pagaza and Othón were twin souls. Both were provincials who held the rapid economic transformation (including urbanization and industrialization) experienced by Mexico in the late nineteenth century in contempt;

FIG. 7.20. Frontispiece to Joaquín Arcadio Pegaza, *Murmurios de la selva* (1887). Modernists like Pegaza abandoned the use of landscapes to articulate historical narratives. Pegaza was a latinist who was influenced by Virgil's pastoral view of the nation, but he was also influenced by romantic German sources.

both saw themselves as defenders of quickly vanishing idealized pastoral ways; and both turned to Virgil and Horace for inspiration, thus reviving long-abandoned neoclassical sensibilities in Mexico. In his "Al volver al campo" (in *Murmurios de la selva* [1887]), Pagaza decried the urban experience that overwhelmed the senses and rendered viewers unable to appreciate the beauties of nature. He repeatedly published poems about the joys of returning home to an idealized Valle de Bravo, his rural birthplace northwest of Toluca. Yet Pagaza was no rural bumpkin, but a sophisticated cosmopolitan. An exquisite latinist who ended his career as bishop of Veracruz, he translated and sought to imitate Virgil, particularly the latter's eclogues (poetic dialogues among shepherds). In an 1887 introduction to Pagaza's poetic essays on the landscape, Rafael Angel de la Peña, secretary of the Mexican Academy of Language, showed that the bard had revived Virgil and Horace because Pagaza had grown tired of the type of landscape narratives so earnestly promoted by the likes of Altamirano, Riva Palacio, and

Velasco. Critics like Altamirano, blinded by a nationalist agenda, thought of the pastoral as "frivolous" and unsuited to the nation's new historical circumstances: audiences needed to be introduced to local, not classical, landscapes. Pagaza, de la Peña argued, believed in absolute standards of beauty, which made Virgil relevant for all ages and countries. His Platonic aesthetics drew upon the writings of nineteenth-century German philosophers for support. Pagaza's pastoral views of the landscape, in fact, emerged out of his engagement with the writings of Kant, Schiller, and Hegel.[53]

Othón was also a bard profoundly influenced by German ideas. Like Pagaza, Othón decried the "ruin and miserable crowdedness of the city; its ramparts and walls, its palaces and temples, and its obelisks, which drown [the senses]."[54] To offset the urban-induced aesthetic blindness to nature's beauty, Othón turned to written descriptions of nature as if it were an impressionist landscape.[55] He sought to do for the written landscape what Wagner had done for music. Like Wagner, Othón turned to the forest for inspiration, seeking to capture the sounds of colors and the colors of sounds in nature. But unlike Wagner's, Othón's forest was a generic one, purposefully detached from any national narrative.[56]

Coda

Having introduced Wagner, it is only appropriate to conclude this study of landscapes and identities in Mexico with a historiographical coda. The material discussed in this chapter simply does not fit the stereotypical image of nineteenth-century Mexico: of seignorial haciendas, rural bandits, corrupt, irresponsible political elites, and disenfranchised Amerindians. The "Mexico" of our historiography is well known and predictable, and historians have found no time to ponder the reception of Wagner among the "elites." In our sociohistorical narratives, such subjects are readily dismissed as irrelevant or, worse, self-deluding strategies on the part of elites whose imitation of foreign fashions is embarrassing. Take, for example, Michael Johns's book *The City of Mexico in the Age of Díaz* (1997). Here the history of the Mexican capital is predictably cast in tragic terms as a city divided between two extremes: the masses of the unwashed (who lived in eastern half of the city, kept their rural ways, and although dismissed as *los pelados* were in fact the "real" and "profound" Mexico) and the select few (the rich and powerful who mindlessly aped French ways and lived in the western half of the capital in their fortified villas). Johns's city is violent, washed in pulque, obsessed with pomp and circumstance, filled with the vacuous rhetoric of laws and elections but ruled through coercion and oppression: "Perhaps no other city in the Americas combined such high levels of mistrust and violence with a penchant for the images of blood and death."[57] The problem with Johns's inordinately tragic account is that the conceit of the "authentic" organizes the narrative. It prompts him to consider the intellectual life of the city as artificial and thus dispensable.

But there is a deeper logic underpinning Johns's tragic narrative of the history of the city. The behavior of the elites is embarrassingly inauthentic because it seeks to ape the "West." This same logic has recently led Peter Gay to write off Latin America entirely in his history of the "Victorian" middle-class experience. Building on the assumption that from Boston to Paris and from Berlin to Moscow, the bourgeoisie shared remarkably similar values, paradoxes, and predicaments, Gay draws upon nineteenth-century sources from Europe, Russia, and the United States, dispensing with historiographies that look at the past along narrow national boundaries. For him, the term "Victorian" encompasses "all of Western civilization, and

[is] synonymous with 'nineteenth century.'"[58] Gay's understanding of what constitutes the "West" is unusually generous, for it includes eastern Europe. And yet Gay has little to say about the distinctly imperial dimension of the Victorian experience, which expanded its tentacles and values from Cairo to India to Japan. Millions of "Victorians" also appeared in Latin America. My problem with Gay's definition of "Western" civilization is that it restricts the boundaries of what was in fact a global cultural phenomenon. Such constructs of the "West" are problematic because they tend to flatten the historical experience of places other than the United States and Europe. Orozco y Berra, who wrote a history of geography to reply to condescending French views of Mexico, would have protested against the exclusion of his fellow Mexican Victorians' richly textured experience from being even so much as a footnote in Gay's account. It is about time to rescue these other Victorians from the condescending historiographical views of the "West."

Notes

1. Chivalric Epistemology and Patriotic Narratives

1. On nineteenth- and early twentieth-century Spanish colonial science in Africa, see González Bueno and Gomis Blanco.

2. For representative syntheses of the history of science in the sixteenth and eighteenth centuries, see López Piñero, *Ciencia y técnica*; Martínez Ruiz; Goodman, *Power and Penury*; Sellés, Peset, and Lafuente. For a survey of eighteenth-century colonial science, see Chapter 3 in this book.

3. Most studies tend to focus on the Spanish scientific expedition to South America of 1862–66; see López-Ocón and Puig-Samper; Puig-Samper; R. Miller; López-Ocón and Pérez-Montes Salmerón. On the development of a Creole Cuban scientific tradition in nineteenth-century Cuba, see Pruna. On the science behind nineteenth-century sugar production in Cuba and Puerto Rico under both Spanish and U.S. imperial rule, see McCook.

4. Schmidt-Nowara.

5. See Varey et al.

6. Robertson; Gruzinski, *Amérique*.

7. De la Cruz and Badiano.

8. On this topic, see Cañizares-Esguerra, "Renaissance Mess(*tizaje*)."

9. On this topic, see Chapter 4 in this book and also Cañizares-Esguerra, *How to Write the History of the New World*.

10. Kagan, "Clio and the Crown."

11. On this, see Armitage, 1–60.

12. De Vos, "Herbal El Dorado" and "Art of Pharmacy."

13. Barrera, "Local Herbs" and *Experiencing Nature*.

14. López Piñero and López Terrada, *Influencia española*; and Schiebinger and Swan. On the history of cochineal trade, see Greenfield. On the history of the vanilla trade, see Ecott.

15. Alvarez Peláez; Mundy.

16. On these traditions, see Sandman; Sala Catalá; Lamb; Bargalló.

17. See figs. 2.4 and 2.5 and related discussion in Chapter 2 in this book.

18. On the thoroughly multinational and multiethnic character of the loosely held Catholic monarchy on the eve of the Bourbon reforms, see Kamen. Alejandro Cañeque (2004) has argued that the Spanish "state" in the New World was no more than the ability of the lay and clerical authorities to muster legitimacy through elaborate (and always highly contested) ritual displays.

19. See Chapter 4 in this book.

20. This idea has been explored in greater detail in Cañizares-Esguerra, *How to Write the History of the New World*, 155–60. The dozens of colonial natural history expeditions financed by the Spanish Crown were not solely efforts to extract riches. There was an important nationalist component to these efforts, including giving Spanish names to the new plants.

21. Lafuente and Mazuecos; Steele; González Bueno; Puerto Sarmiento, *Ilusión quebrada*; Lozoya; Engstrand; Pimentel, *Física de la monarquía*; Frías Núñez.

22. Chapter 6 in this book goes into greater detail about this.

2. THE COLONIAL IBERIAN ROOTS OF THE SCIENTIFIC REVOLUTION

1. Charles V adopted the columnar "plus ultra" device in the 1510s to signal a break with medieval readings of Hercules that emphasized cautious (military) prudence ("non plus ultra"). The pillars and the motto originally meant to signify the willingness of Charles V to launch a crusade through North Africa by crossing the straits of Gibraltar. By the mid sixteenth century, however, the motto took on the meaning of daring transatlantic imperial expansion. On this topic, see Rosenthal, "Invention of the columnar Device" and "Plus Ultra." Tanner, in *The Last Descendant of Aeneas*, argues that the legend of the Golden Fleece was employed in classical and medieval discourses to legitimate emperors beginning with Virgil's *Aeneid* and Eclogue IV (to honor Augustus). The motif was Christianized under Constantine, Clovis, and Charlemagne. In medieval epics, the Fleece came to stand for the recovery of Jerusalem from Islam.

2. On this characteristic of the Spanish Renaissance, see Nader.

3. Maravall, *Antiguos y modernos*, 431–53, 483–549. Hooykaas (*Humanism and the Voyages of Discovery*) has also explored the self-awareness of modernity for the case of the sixteenth-century Portuguese.

4. Camões, 20. Alonso de Ercilla y Zúñiga's *La Araucana* came out in three parts, in 1569, 1578, and 1589. David Quint, *Epic and Empire*, has written a remarkable study that locates Camões's and Ercilla's poems within the classical epic tradition of Virgil's *Aeneid* and Lucan's *Pharsalia*, but he overlooks the self-confident modernity of the sixteenth-century Iberians as a critical configuring factor of their poetry.

5. Lupher, ch. 1.

6. García de Céspedes. To my knowledge, Juan Pimentel was the first to call attention to the similarities (and differences) between these two frontispieces. See Pimentel, "Iberian Vision."

7. Hakluyt, 1: *3r–v and 4v. Hakluyt most likely was referring to Alonso de Chaves's "Quatri partitu en cosmogarphia practica, y por otro nombre, Espejo de navegantes" (MS written in the 1530s that remained unpublished but circulated widely); Jerónimo de Chaves's translation of Johannes Sacro Bosco's *Sphera mundis, Tractado de la sphera* (Seville, 1545); one of the multiple editions of Chaves's *Chronographia* (1548, 1561, 1566, 1572, 1581, 1584); and Rodrigo Zamorano's *Compendio del Arte de Navegar* (Seville, 1581), *Cronología y repertorio de la razón de los tiempos* (Seville, 1594), and *Los seis libros de geometría de Euclides* (Salamanca, 1576). Surprisingly,

Hakluyt does not mention Pedro de Medina's *Arte de navegar* (Valladolid, 1545) and *Regimiento de navegación* (Seville, 1563), which were translated and reprinted several times in England.

8. Barrera, "Local Herbs" and *Experiencing Nature*.

9. On the influence of the chivalric discourse on early-modern English science, particularly mathematics, see Alexander. Alexander, however, does not trace these influences back to Spain and Portugal. In light of all the criticism of the cruelty of the Spanish conquest by the English, especially in the Elizabethan period, it might appear counterintuitive to suggest that Bacon considered the Spanish colonies a model. Just when Bacon was writing his *New Atlantis*, however, the Virginia Company held up the Spanish colonies as an example to follow in every single respect. Having faced the loss of a third of the settlers of Virginia to a rebellion of indigenous Tidewater peoples, the official spokesman of the company, Edward Waterhouse, argued that the massacre had occurred only because the English had failed to follow in the footsteps of Spain, and that they needed to learn how to do things right from the likes of Cortés and Pizarro. The natives of Virginia should be conquered and enslaved by a policy of divide and rule. Moreover, Waterhouse praised the Spanish colonies for what "industry, patience and constancy" could accomplish. Arguing against the grain of what would later become centuries of accumulated wisdom about Spanish colonialism, Waterhouse said that to succeed in the New World, the English had to learn from the Spaniards to stop looking for treasure and turn to the production of agricultural staples and export commodities. See Waterhouse, 30, 32–33.

10. Goodman, *Power and Penury*, 72–73.

11. Pimentel, *Testigos*, 73–94.

12. Baroni Vannucci.

13. On the "discovery" of the New World as the discovery of new stars and constellations, see Chapter 6 in this book.

14. Kagan, "*Arcana Imperii.*" For a wonderful study of early modern Iberian scribal culture in the age of the printing press, see Bouza.

15. For recent examples of new interest on Iberian natural history, see Schiebinger and Swan; Smith and Findlen; Ishikawa.

16. Bonneville, 61.

17. La Porte, 16: 94.

18. Quoted in Tietz, 100.

19. Hillgarth; Juderías; García Carcel; Maltby; Schmidt.

20. Montesquieu, 115.

21. Daston and Park, 146–49. See also Nieremberg, *Curiosa filosfía* and *Ocvlta filosfía*.

22. MacCormack; Cervantes, 24–31.

23. Clark, esp. pt. 2; Cañizares-Esguerra, *Puritan Conquistadors*, ch. 4

24. José de Acosta, *De Christo*.

25. See also Prest.

26. Drayton, 13.

27. Monardes, *Dos libros*; Orta; Cristóbal Acosta.

28. Clusius, *Aromentum et simplicium aliquot medicamentorum*; *Simpliciis me-*

dicamentis; and esp. *Rariorum aliquot stirpium per Hispanias* (1576), in which he repeatedly cites botanical gardens he visited or friends from whose botanical gardens he obtained samples. The network was vast, including contacts with gardens in Belgium, France, Austria, Italy, and England. Among the botanical gardens in Portugal and Spain he visited were those of the Divae Virginis monastery on the outskirts of Valencia (16, 444); Ferdinand Cotinho, a Portuguese gentleman (131, 280); Johannes Plaza, a "most learned" physician in Valencia (254, 289, 444, 479); the royal palace in Lisbon (299); and Pedro Alemán (444).

29. Clusius, *Aliqvot notae.*

30. Clusius, *Admiranda narratio.*

31. See Dandelet.

32. Iacobus Mascardus Typographus Lectoris, in Hernández, *Rerum medicarum Novae Hispaniae thesaurus,* n.p.: "Si Herbariae, si naturali Historiae addictus: si Medicus, si Philologus, si denique Phytosophus: si Florilegus Principium, aut Heroinorum blandiris deliciis: vel si novis e fructibus, aut Pharmacis Mercator, Institor, Pharmacopola, Odorarius, sanitatem, oblectamenta quaeris, aut lucrum, novas hic certe habes Rerurm, Imaginum, Vocabulorum Myriades, quae & oculos, & mentem, & desiderium omni ex parte explere possint."

33. Liber Tertius, Arbores describit, in Hernández, *Rerum medicarum Novae Hispaniae thesaurus,* 44: "Ne dicam quicquam de ingentibus illis arborum trabibus, quibus domicilia nobis paramus, quo tutiores a ferarum incursionibus, & tempestatum inundationibus persistamus, & naves aedificamus, quibus nos nostraq. committentes audacissimi homines turbulentissimo, & minacissimo oceano fidimus ignotas Terras, ignotaq hominum commercia exquirimus."

34. See Dandelet.

35. De La Cruz and Badiano. Freedberg misidentifies Martín de la Cruz as Juan. He also presents Badiano as a humble "Indian" given to self-deprecating remarks, compounding the stereotype of exploited Amerindians and haughty Spaniards. Clearly, Freedberg is unaware of the rhetorical and cultural conventions of classical Nahuatl (263–64). The Codex, originally entitled *Libellus de medicinalibus Indorum herbis,* in more ways than one resembles the images and text in Hernández's original manuscript, clearly showing that Hernández's was a collective work in which Nahua intellectuals played a central role.

36. The commentaries by Johannes Faber on some of Hernández's illustrations of animals, one of the longest sections of the *Rerum medicarum Novae Hispaniae thesaurus* (457–840), depend utterly on information Faber obtained in Rome and Naples from Spanish and Spanish-American intellectuals. Among those Faber relied on the most was Gregorio de Bolívar, a Spanish Franciscan who spent twenty-five years in the remotest parts of Peru and Mexico, as well as additional years in the East Indies, and who was very learned in three indigenous languages. Throughout, Faber copied verbatim page after page of Bolívar's unpublished treatise on the natural history of the New World (506). Faber also drew on the work of Bernardino de Córdoba, a grandee addicted to the study of nature who assembled a collection of exotic things and animals from the Indies in Naples (550). Pedro de Aloaysa, a Dominican born in Lima representing the American interests of his order in Rome, educated Faber about all things American in long, friendly conversations (695). Finally, Bartolomé

de la Ygarza, a Spanish Dominican who had lived in America for seven years, willingly transmitted his knowledge of American animals and plants to Faber (743).

37. Mestre, 53–81; François Lopez, "Comment l'Espagne eclairée inventa le Siècle d'Or."

38. Goodman, "Scientific Revolution"; Romano.

39. The procedure was banned, however, because the Crown captured a sizeable amount of the wealth produced in the mines by reason of its monopoly on mercury.

40. There are, to be sure, more detailed studies of the history of amalgamation and other technological improvements in the silver mines of colonial Spanish America; see, e.g., Menes-Llaguno; Bargalló; Muro; Bakewell.

41. Newman; Newman and Principe.

42. López Piñero, *Ciencia y técnica*, 34–37.

43. Masdeu; Lampillas; Andrés.

44. Navarro Brotóns, "Reception of Copernicus"; Navarro Brotóns and Rodríguez Galdeano.

45. See Lamb. Building on Lamb's scholarship, Alison Sandman (2002) has recently offered a provocative interpretation of the reason why academies to train learned ships' pilots found courtly support in sixteenth-century Spain. A seamanship obsessed with locating latitudes and longitudes on maps was largely the result of competing Portuguese and Spanish imperial claims to Asian and American territories.

46. Navarro Brotóns, "Reception of Copernicus."

47. Muñoz, *Libro del nuevo cometa* (Valencia: Pedro de Huete, 1573). Navarro Brotóns has published this rare book along with other writings by Muñoz (Valencia: Valencia Cultural, 1981).

48. Navarro Brotóns and Rodríguez Galdeano.

49. For a survey of recent scholarship, see Martínez Ruiz, the proceedings of an international congress on science and technology under Philip II.

50. For example, an entire issue of a leading Spanish journal on the history of science and technology, *Arbor* 152, 604–5 (April–May 1996), was devoted solely to assessing the impact of López Piñero's *Ciencia y técnica*.

51. Navarro Brotóns, "Ciencias."

52. On these traditions, see Albuquerque; Seed, *Ceremonies of Possession*, 107–16; Corteseão.

53. López Piñero and Calero; Pardo Tomás and López Terrada; Fresquet Febrer; López Piñero and López Terrada.

54. Pomar; López Piñero and Pardo Tomás, *Nuevos materiales* and *Influencia*; Bustamante, "De la naturaleza," "La empresa," and "Francisco Hernández."

55. MacDougall.

56. Mukerji; Coffin, *Villa in the Life of Renaissance Rome.*

57. Prest.

58. Strong; Comito.

59. Lazzaro, esp. chs. 7 and 9; Coffin, *Villa d'Este.*

60. Darnall and Weil; Strong; Comito.

61. Añón Feliú, "La Granja," "Nature," "Immagine della natura."

62. Alvarez Peláez, 152–53.

63. Sala Catalá, 16.

64. For a marvelous study of colonial Spanish American urbanism, see Kagan, *Urban Images.*

65. For an interpretation of the early modern Spanish empire as a vast, global, multinational, and multiethnic array of actors awkwardly (yet efficiently) brought together by the Crown, see Kamen.

3. FROM BAROQUE TO MODERN COLONIAL SCIENCE

1. I am limiting my remarks here to the elite systems of natural philosophy that disappeared after the Spanish conquest. Indigenous scientific knowledge has survived to this day in the vernacular tradition.

2. Farriss; but cf. Clendinnen; Lockhart. Serge Gruzinski has described the rise and fall of a hybrid Amerindian-Christian culture in the central valley of Mexico in the mid sixteenth century. Observant Franciscan friars, using techniques derived from Renaissance humanism, trained a cadre of native classicists, who acted as cultural translators. Native and European humanist friars produced polyglot texts (Latin, Spanish, and Nahuatl), including a monumental encyclopedia of Nahua lore, the Florentine Codex, and a Nahua herbal, the *Libellus de medicinalibus Indorum herbis* (or Codex Badianus). The *Libellus* introduced Nahua glyphs and esthetic conventions into the genre of European herbals. See Mundy; Gruzinski, *Colonisation de l'imaginaire*, ch. 1, esp. 76–100.

3. Craig A. Russell, personal communication. "Standing" means a fixed body of instrumentalists rather than individuals hired ad hoc, as seems to have been the case in seventeenth-century court operas and ballets. This new arrangement led to new technical developments in musical style, from baroque counterpoint to classical style. Mannheim's opera dates to the 1720s.

4. Lanning, *Academic Culture.*

5. Martín, 97–118. See also Harris.

6. Philip III appointed the Jesuits sole official cosmographers in 1628, consolidating the teaching of astronomy at the Jesuit Imperial College in Madrid. See "Expediente sobre la asignación de la cátedra de cosmógrafo en el Colegio Imperial de Madrid," with documents dating from August through October 1760, AGI, Indiferente General 1510.

7. On coordinated astronomical observation by Jesuit cartographers, see Christian Reiger, "Memorial del cosmógrafo mayor al Consejo de Indias sobre limitaciones que tiene el cosmógrafo para ejercer las expectativas puestas sobre el deacuerdo al título," June 30, 1761, AGI, Indiferente General 1520.

8. Leonard, *Don Carlos*; Trabulse, "Obra científica de Don Carlos de Sigüenza." Sigüenza in many ways resembled other European baroque polymaths; see Eriksson.

9. Maravall, *Cultura del barroco*; Leonard, *Baroque Times.*

10. Paz, *Sor Juana.*

11. This correspondence can be found in Osorio Romero, *Luz imaginaria.* On Kircher in Spanish America, see Findlen, pt. 5.

12. Kircher, *Magneticum.* Osorio has also reproduced Kircher's Latin dedication

as well as excerpts from Kircher's *Magneticum* with references to the work and contributions of Favián to the study of magnetism. See Osorio Romero, *Luz imaginaria*, 111–28.

13. "Testamento de Don Carlos de Sigüenza y Góngora," in Pérez de Salazar, 170–72.

14. Brading, *First America*.

15. For examples of micro- and macrocosmic analogies, see Osorio y Peralta. On emblematic-religious reading of natural objects, see Calancha, *Corónica moralizada*, 57–59 (the granadilla plant and flower resemble the symbols of the Passion—nails, sponge and lance, wounds, bindings, and crown of thorns—and therefore induce pain and general malaise); and Vetancurt, 1: 22–23 (precious stones reveal their medicinal value through the symbolism of their colors: white ones cure milk-related illnesses; red ones cure blood diseases; green ones with black spots stop bilious-hepatic attacks [*hijadas*]; green ones with red spots cure intestinal bleeding; and, finally, green ones with white spots help dissipate kidney stones), 38 (a spring turns out to be medicinal because it produces crosslike stones), 42 (bananas whose cores look like a crucified Christ), and 51 (*tlahulitucan* trees whose crosslike leaves keep demons away). On patriotic astrology, see Chapter 4 in this book.

16. Calancha, *Corónica moralizada*, 48–50, 58–59.

17. Sánchez, *Imagen de la Virgen*, 168 (crown of twelve stars); 119 (eclipsed sun); 223–24 (moon); 226–27 (forty-six stars).

18. Brading, *First America*, ch. 16; Lafaye. See also Cañizares-Esguerra, *Puritan Conquistadors*, ch. 2

19. Dozens of treatises appeared in the seventeenth and eighteenth centuries seeking to uncover the hidden meanings of the image. For example, some scholars thought that an eightlike figure on the robe of the Virgin signified that the image was the eighth wonder of the world. Others thought that the mark was in fact a Syrian-Chaldean character that, along with other "Oriental" symbols in the picture, indicated that St. Thomas had come to Mexico with the image in the first century A.D. after having preached the gospel in the Orient. See Fernández de Echeverría y Veytia, 12; Cabrera, 519; Borunda, 276–77; Mier, 1: 249–50.

20. Maza, *Guadalupanismo mexicano*, 43–45, 177.

21. Florencia, 394–95. See also Eguiara y Eguren, 487; Cabrera y Quintero; Fernández de Echeverría y Veytia

22. Trabulse, *Ciencia perdida*.

23. Diego Rodríguez, fol. 4v. The more "modern" parts of Rodríguez's *Discurso etheorológico* are reproduced in Trabulse, *Historia de la ciencia*, 324–37.

24. Farías.

25. Francisco Fuentes y Carrión, "Discurso sobre las virtudes del pulque" (1733), Biblioteca Nacional de México, MS 1540.

26. Cayetano Francisco de Torres, "Virtudes maravillosas del pulque, medicamento universal o polychresto" (1748), Biblioteca Nacional de México, MS 13, fols. 1–16. Colonial physicians used etymology to identify medical virtues. In a heated debate that took place in Mexico City in 1782 over the efficacy of live lizards for curing tumors, the *protomédico* (a learned physician charged by the Crown with the regulation of medical practice in the city), Joseph Giral Matienzo, supported José Vicente

García de la Vega, who claimed that lizards could cure not only cancer but many other illnesses as well, including drawing out splinters by means of hidden sympathies. Giral Matienzo argued that García de la Vega had understood that lizards were "robust" remedies, able to cure many diseases, because he had identified the meaning of their "hieroglyph," namely, the hidden Latin etymology for lizard (*lacertus*, a synonym for *robur*, robust). See Joseph Giral Matienzo, "Aprovación," in García de la Vega, n.p.

27. Hernández, *Antigüedades*, 118, 147.

28. Libro de Claustros, University of San Marcos, 1637, quoted in Unanue, "Introducción," 71.

29. José Eusebio de Llano Zapata, letter to the marquis of Villa Orellana, ca. 1761, in Llano Zapata, 595. Llano Zapata proposes the creation of a "College of Metallurgy" to train students in experimental philosophy, mathematics, geometry, hydraulics, mechanics, and natural history, as well as in Italian, French, German, Greek, Latin, and Quechua. Scholarship on Llano Zapata has received a boost with the discovery and publication (2005) of many of his long-lost manuscripts, including two of the three volumes of his *Memorias históricas, físicas, critico, apologéticas* and his *Epítome cronológico o idea general del Perú* (1776).

30. Bermúdez, 178–79.

31. Caldas, "Prefación a la geografía de las plantas de Humboldt," in *Obras completas*; Moreno, 25.

32. Martín de Sessé, letter to Casimiro Gómez Ortega, January 15, 1785, reproduced in Lozoya, 30.

33. Herr.

34. On the rise of a new cultural authority for science linked to the search for new forms of political legitimacy, see Outram, 47–61, 96–113. This pattern of subordination of colonial science to mercantilist policies has also been identified in eighteenth-century Haiti by McClellan.

35. Capel, *Geografía y matemáticas*. The largest and most significant of these cartographic expeditions was that headed by Alejandro Malaspina, which also included numerous naturalists and painters. See Engstrand; González Claverán; Pimentel, *Física de la monarquía*.

36. For a survey of recent historiography, see Puig-Samper and Pelayo, "Expediciones botánicas." The literature on eighteenth-century botanical expeditions has witnessed an explosion. Some representative titles are Steele; Lozoya; Frías Núñez.

37. Puerto Sarmiento, *Ciencia de cámara*, 148–209; Frías Núñez, 159–250.

38. Puerto Sarmiento, *Ciencia de cámara*, 174. When compared to the colonial botanical agendas of other European powers, the Spanish emphasis on the search for pharmaceuticals in the tropics appears exaggerated. On the eighteenth-century colonial botanical agendas of Britain and Sweden, see Miller and Reill. On France, see McClellan, 111–15, 147–61.

39. Whitaker; Bargalló.

40. Italian and French court physicians in the entourage of the Bourbons also played an important role in Spain's scientific renewal; see Riera. Italian and French doctors arrived in the colonies as court physicians for viceroys and prelates, introducing Newtonian and iatromechanical ideas early in the eighteenth century; see,

e.g., Bottoni; Beaumont. Federico Bottoni, who arrived in Spain in the entourage of Isabel de Farnecio, second wife of Philip V, and worked in Lima as court physician for two viceroys and for a Franciscan prelate, introduced Peruvian doctors to Harvey's theory of circulation. Juan Blas Beaumont, a surgeon and latinist, holder of the chair of anatomy at the University of Mexico, and in the retinue of the archbishop Francisco Antonio de Lorenzana, was in all likelihood the son of Blas Beaumont, a French surgeon at the court of Philip V who contributed to the renovation of early eighteenth-century Spanish medicine.

41. John Tate Lanning has sought to give Spanish American colonial universities their due by claiming that they played a key role in the eighteenth-century cultural renewal; see Lanning, *Eighteenth-Century Enlightenment*. Yet the evidence against Lanning's thesis seems to be overwhelming; see, e.g., Enrique González; Baldó; Ten; Soto Arango, "Enseñanza ilustrada." For a brief overview of the resistance of Spanish universities to change, see Peset and Peset.

42. See also Capel, Sánchez, and Moncada.

43. The Academy of Art of San Carlos was founded in Mexico in 1783 to teach artisans and masons optics and mathematics, among other things; see Thomas Brown.

44. For current scholarship on the new Spanish scientific institutions described in this paragraph, see the articles in Lafuente and Sala Catalá.

45. Puerto Sarmiento, *Ilusión quebrada*.

46. Kendall Brown; Contreras and Mira.

47. Saldaña, esp. 43–46.

48. Lanning, *Royal Protomedicato*.

49. On the public sphere in the colonies, see Rodríguez O., *Independencia*, 58–63 and passim; Silva; Clement, "Apparition de la presse."

50. Cígala. With the approval of the dean of the University of Mexico, Juan José Eguiara y Eguren, and of a leading Mexican Jesuit, Francisco X. Lazcano, Cígala took Feijoó, a Benedictine friar largely responsible for the popularization of Newton and Descartes in early eighteenth-century Spain, to task for challenging Aristotle and scholastic theology. Cígala, Eguiara y Eguren, and Lazcano, however, were not blindly holding to the past. They chastised the moderns for claiming to have created a new philosophy when it had already been developed by the ancients. Cígala criticized Feijoó from a position of strength, revealing Feijoó's lack of understanding of the mechanics of air and of the writings of Boyle and Leibniz. For an unsympathetic reading of Cígala as throwback reactionary, see González Casanova, 114–29. On Newton in colonial Spanish America, see Arboleda, and Lertora Mendoza.

51. *Gazetas de México* 4, 16 (August 1790): 152–54, describes at least ten cabinets of experimental apparatus in Mexico City by that time.

52. Francisco Xavier Alexo de Orrio, "Metalogía o physica de los metales," Biblioteca Nacional de México, MS 1546; Andrés Ibarra Salazán (AYS), "Tratado de las montañas y rocas" (written ca. 1810), Biblioteca Nacional de México, MS 1510. Ibarra Salazán, however, compares geological layers to layers of tissue of the human body (see fols. 15v–16r).

53. According to Mary Terrall, the expeditions to measure a degree of the meridian

were not solely about settling the Cartesian-Newtonian debate but, more important, about rivalries between Pierre-Louis Moreau de Maupertuis and the Cassinis (Jean-Dominique and Jacques) over which instruments and techniques of measurement to use in mapmaking. Maupertuis's success in an expedition to Lapland settled the dispute in his favor without having to wait for La Condamine to arrive. See Terrall.

54. La Condamine, *Journal du voyage*, passim. The events surrounding Seniergues's murder are exquisitely recounted by La Condamine in *Lettre à Madame****. Two other French expeditions had already visited Peru prior to La Condamine's, suggesting an early pattern of Spanish-French collaboration under the Bourbons that lasted throughout the eighteenth century. See Feuillée, *Journal* (1714; 1725); Frézier. Hamy recounts yet another French expedition to the Andes led by Joseph Dombey in the last quarter of the century. See also the expedition of Jean-Baptiste Chappe d'Auteroche.

55. Humboldt, *Voyage de Humboldt et Bonpland*.

56. Glick.

57. Cañizares-Esguerra, *How to Write the History of the New World*, 28–84.

58. Cañizares-Esguerra, "Utopía de Hipólito Unanue."

4. NEW WORLD, NEW STARS

1. Salinas y Córdova, *Memorial, informe, y manifiesto*, fols. 18v–19r (on America as microcosm); 17v (America as location of paradise due to its climate and peoples); 22v–23r (Amerindians and Creoles); 37 (Amerindians suited to create ideal Christian communities).

2. Ibid., fols. 11v, 106v.

3. Salinas y Córdova, *Memorial de las historias* (1630), ch. 1. See also 1957 ed., 11–12.

4. Morner; Israel; Seed, "Social Dimensions"; Cahill.

5. León Pinelo, *Paraíso*, 2: 524–29.

6. Ibid., 2: 4–6.

7. Joyce E. Chaplin has made a similar argument. Although the English colonists articulated a view of their bodies as better adapted to the American climate than those of the Amerindians, who were dying by the thousands due to exposure to new European diseases, it appears that the colonists were unable to turn their new-found exceptional bodily self-identification into a full-fledged racialist discourse. According to Chaplin, they remained bound to the assumption that environment was the ultimate cause of bodily transformations. My argument is that the Spanish American Creoles were far more successful at transforming a racial "idiom" into a full-fledged racialist discourse than their English counterparts.

8. Popkin. On sixteenth- and seventeenth-century debates on the origins of Amerindians, see Huddleston.

9. Their arguments paralleled those medieval discourses that through the doctrine of the curse of Noah sought to separate the bodies of peasant commoners from the nobility's. On this, see Freedman.

10. Elliott, *Old World*. See also Pagden, *Fall*; Grafton, *New Worlds*; and the articles in Kupperman, ed., *America*. On the classical tradition of environmental determinism, see Glacken.

11. Stuart Hall.

12. Stoler, 16.

13. Cañizares Esguerra, *How to Write the History of the New World.*

14. On the nature and complexity of early modern science, see Lindberg and Westman.

15. Allen; Geneva; Smoller. Siraisi's *Clock* and Grafton's *Cardano's Cosmos* make it perfectly clear that the principle of astral effects on the human body was one of the pillars of early modern medicine. On medieval and Renaissance theories of temperaments and constitutions, see Siraisi, *Medieval and Early Renaissance Medicine,* 97–114.

16. Fernández de Oviedo, *Sumario,* 77–78.

17. As early as 1519, the Spanish cosmographer Martin Fernández de Enciso attributed the great frequency of lightning storms in the Caribbean to the greater humidity of the Indies; see Fernández de Enciso, fols. 8v–9r.

18. Cardano quoted in Anghiera, *Decades of the newe worlde,* trans. Eden, 184r–v. On Cardano, see Siraisi, *Clock*; Grafton, *Cardano's Cosmos.*

19. José de Acosta, *Historia* (1591), bk. 2, ch. 6, 58v–59r.

20. Ibid., bk. 3, ch. 28, 123v.

21. Valadés, pars. 2, ch. 27, 94 [228], "sint stupidi, ta[m]q[ue]; in crasso aëre nati."

22. Cárdenas, 37–42, 231–32. As Cárdenas explained it, humidity sapped the body of its vital heat, paradoxically causing it to desiccate and age more quickly.

23. Torquemada (1615), bk. 14, ch. 19, 613 (the climate of the New World made food less nourishing and weakened the body). The first reference to the nonnutritious nature of the American food was voiced by Anghiera, *De orbe novo,* decade 1, ch. 10, dvi (verso). In the seventeenth century, these misgivings continued. About 1650, a Spanish naturalist long resident in the Indies, the Jesuit Bernabé Cobo, argued that the humidity of the Indies was a compensating mechanism to keep the continent temperate but also the reason why fruits there were cold, humid, nonnutritive, and even insalubrious; see Cobo, *Obras* 1: 55–56, 237. In 1698, the Franciscan Creole Agustín Vetancurt maintained that the temperate climate of the Indies resulted in shallow roots (in cold climates, roots went deeper) and, therefore, caused fruits and staples to be nonnutritive. According to Vetancurt, the nonnutritive quality of food, however, had an advantage for European colonists, because the less substantial the meal, the fewer the internal vapors created inside the body and the brain. Since colonists, therefore, had fewer vapors clouding perceptions, it could be argued that they were more intelligent than Europeans. See Vetancurt, treatise 1, ch. 6, 10–11. The assumption that American food lacked substance lasted well into the eighteenth century. In 1751, the Creole doctor Joseph Francisco de Malpica Diosdado wrote a treatise encouraging fasting. In the New World, he argued, doctors had long excused the population from fasting because they assumed that the climate made the local population weaker and the food less nutritious. Malpica Diosdado maintained that such views gave ammunition to the Pre-Adamites by emphasizing the radical difference between the organic productions of the Old and New Worlds. See Malpica Diosdado, esp. 151–65.

24. Gage, 42–43.

25. On this tradition, see Maclean; Laqueur; Schiebinger, ch. 3. Laqueur and

Schiebinger argue that up until the late eighteenth century, European scholars did not see men's and women' bodies as ontologically different. Female bodies were thought to be male bodies manqué, whose sexual organs had been inverted due to the lack of vital heat and excess humidity. Thus it was common to think that spontaneous sexual transformations were not only possible but happened often. Peter Brown (10–11) has argued that male Roman elites educated in the Galenic classical medical tradition saw their bodies as constantly threatened by physiological processes that led to the depletion of vital bodily heat (sexual intercourse) and thus to emasculation.

26. Suárez de Peralta, 5–6, and Conquistador anónimo, 68. Views of "savages" and racial inferiors as feminized others have enjoyed widespread popularity in the West; see Stepan, "Race and Gender."

27. On perceptions of the American climate in British America, see Kupperman, "Puzzle," "Fear of Hot Climates," and "Climate and Mastery." Thanks to Professor Kupperman for making copies of these three articles available to me.

28. Tooley; Glacken.

29. Although the fifteenth-century Portuguese expansion to Africa was predicated on exquisite astronomical knowledge of the "new stars and skies" of the Southern Hemisphere, this new knowledge was not made readily available. See Seed, Ceremonies, 100–148. According to Seed, 104–5, Master John, a Portuguese pilot, was the first to offer an accurate drawing of the Southern Cross in 1500; it is not clear, however, when this report reached the Crown and whether it was made public.

30. Vespucci, 52–53.

31. Ibid., 35, 40, 90.

32. Ibid., 66.

33. Anghiera, Decades of the newe worlde, trans. Eden, 247v–248r.

34. Fernández de Oviedo, Sumario, 94.

35. Fernández de Oviedo, Historia, bk. 2, ch. 11. Anghiera refers to the testimony of the two Pinzón brothers, who on an expedition to the New World in 1499 found the southern skies unfamiliar, with many unknown constellations. See Anghiera, De orbe novo, decade 1, ch. 9, n.p. It seems curious that the Spanish Habsburgs appear not to have incorporated the new constellations of the Southern Hemisphere into courtly rituals of legitimization as Mario Biagioli has described for the Medici's use of Galileo's new celestial discoveries (i.e., the "Medicean Stars"). On Spain's distinct early modern court culture, see Elliott, Spain, pt. 3.

36. Anghiera, Decades of the newe worlde, trans. Eden, title page.

37. Richard Eden, preface to his translation of Martín Cortés's Arte de navegar (1561), in Arber, xlii.

38. Anghiera, Decades of the newe worlde, trans. Eden, fols. 221r; 244v–248r; 321v–322r.

39. Ibid., 206r–v.

40. Scaliger, 141v: "Ibi tamen auri parum. Pauciores gemmae. Victus ferinus belluinus. Aromatum species, not multa, non nobiles, non optimae." On Cardano, see Siraisi, Clock; Grafton, Cardano's Cosmos.

41. Sahagún, Historia, 3: 157–68. On Franciscan millenarianism in sixteenth-century Mexico, see Phelan, Millennial Kingdom.

42. Hernández, *Antigüedades*, 46.

43. Ibid., 97, 101. It was a commonplace of Spanish colonialism that ancient Amerindian rulers had understood their vassals' nature and the environment and had thus known that their subjects needed to be forcibly made to work, and the learned often urged the Spanish authorities to imitate them in this. See, e.g., Tomás López Medel a los Reyes de Bohemia, Guatemala, March 25, 1555, Archivo General de Indias, Guatemala, 9A.

44. Sahagún's thesis did not reach the European public until the nineteenth century. Parts of Hernández's work on Amerindian antiquities were published by Nieremberg in *Historia naturae*. On the fate of Hernández's manuscripts, see Chapter 2 above.

45. José de Acosta, *Historia*, 1591 ed., bk. 1, ch. 5: 17v (on smaller stars); bk. 1, ch. 2, 13r (on darker spots in the Milky Way).

46. Herrera y Tordesillas, 1601–15 ed., decade 1, bk. 1, ch. 5, 10–11. See also 1934–57 ed., 2: 27–36.

47. Botero, 200 (quotation).

48. Ibid., 199.

49. De la Puente, bk. 3, ch. 3, 21.

50. Quoted in Brading, *First America*, 363.

51. Purchas, 911–12.

52. León Pinelo, *Epitome*.

53. Richard Eden, preface to his translation of Martín Cortés's *Arte de navegar* (1561), in Arber, xlii.

54. García, *Origen* (1981), bk. 2, ch. 5, § 3, 73. The tradition of considering the Carthaginians to be the original ancestors of the Amerindians might, however, have had its origins in the ancient Roman view of the Carthaginians as malevolent savages, given that almost all our information about the Carthaginians comes from Roman sources. I am thankful to Peter Dreyer for calling Roman detestation and denigration of the Carthaginians to my attention.

55. Ibid., ch. 4, § 2, 56–57, § 6, 63, and § 7, 64; ch. 5, 68–69.

56. Ibid., ch. 5, § 1, 70.

57. Lizana, 1988 ed., 54.

58. Rocha, 69, 212–17. Notice that García, Lizana, and Rocha were postulating theories of speciation that resembled Charles Darwin's, particularly Darwin's emphasis on the role played by isolation.

59. On Creole patriotism, see Brading, *First America*; Rubial García; Lavallé; Lafaye; Pagden, *Spanish Imperialism*, chs. 4–5; Phelan, "Neo-Aztecism."

60. Brading, *First America*, 293–301. Rubial García; Israel, 84–87. Israel describes how Archbishop Juan Pérez de la Serna ordered the imprisonment of the Spanish Jesuit Gómez after the latter had denounced Creoles as incompetent and corrupt and incapable of holding public office in a sermon in August 1618. The archbishop also ordered sermons praising the Creole intellect.

61. Leonard, *Baroque Times*, 85–98; Trabulse, *Círculo roto*, 34–37.

62. The inquisitorial documents upon which both Trabulse and Leonard base their generalizations have been published. See, e.g., Quintana, 193–94.

63. Mínguez, *Reyes solares* and *Reyes distantes*; Trabulse, *Círculo roto*, 30; Paz.

For a characterization of baroque science, including astrology, in colonial Spanish America, see ch. 3 above.

64. Calancha, 48–49.

65. Ibid., 239–42 (Lima); 486–87 (Trujillo); 523 (Chuquisaca); 548–49 (Pascamayo); 747 (Potosí); 866–67 (Pucarani).

66. Ibid., 49–50 (on Cruzero); 56–59 (on crosslike plants and fossils). For more examples, see ch. 3 above.

67. Ovalle, bk. 1, ch. 22, 49–51. Ovalle overcounted the number of stars in Volans (as 7, whereas Keyser and Houtman had counted only 5), in Dorado (5 instead of 4), in Toucan (8 instead of 6), in Phoenix (14 instead of 13), in Crane (13 instead of 12), in Indus (12 instead of 11), in Apus (12 instead of 9), and in Triangulum (5 instead of 4). However, he lowered the counts of the Southern Cross (4 instead of 5); Chameleon (5 instead of 9); and Pavo (16 instead of 19). He also included a chart of the southern skies, a slight variation on that offered by Corsali some hundred years before. On Keyser and Houtman and the popularization of their catalogues, see Warner, 14–16, 18–19, 28–31, 121, 201–6. For a translated copy of Houtman's catalogue, see Knobel. At about the same time that Ovalle published his natural history of Chile, the Jesuit Bernabé Cobo, a Spaniard who spent most of his life in Peru and Mexico, articulated a similar defense of the southern skies. Cobo appears not to have known the catalogues of Keyser and Houtman. However, he described a series of southern constellations (Cruzero, Triangulum, and a third one to which he gave the idiosyncratic name Fiducia, which appears to correspond with Hydra). Cobo also described two Magellanic Clouds. He asserted that the Southern Hemisphere had bigger and more numerous stars than the Northern, and in particular that the Milky Way was more luminous and had more stars in the Southern Hemisphere. See Cobo, *Obras*, 1: 27–31.

68. Diego Rodríguez, fol. 4v. Trabulse has reproduced parts of Rodríguez's *Discurso* in *Historia*, 324–37.

69. Diego Rodríguez, 24r–32v.

70. Guamán Poma de Ayala, 1: 40.

71. Las Casas, *Apologética*, 1: 1–205, esp. 158–63, 169–74, 201–5. Las Casas's only negative observation was that the natives tended to reproduce when they were young and their "seeds" were still "humid." Such humidity caused their offspring to have some vapor in their brains and therefore some perceptual distortions. On medieval faculty psychology and later developments in the Renaissance, see Steneck, *Science*, 130–37, and "Albert on the Psychology of Sense Perception"; Park.

72. Gibson, *Aztecs*, 117–18; Stern, *Peru's Indian Peoples*, 102–4, 156–57.

73. On the "myth of the lazy native" in other colonial settings, see Alatas.

74. Lope de Atienza, "Compendio historial del estado de los indios del Piru, con mucha doctrina y cosas notables de ritos, costumbres, e inclinaciones que tienen, con otra doctrina y avisos para los que viven entre los neophitos," Archivo de la Real Academia de Historia, Madrid, 9/4790, Colección Muñoz, vol. 11, fol. 25r.

75. Ibid., fols. 37r and 56r–v.

76. Solórzano Pereira, bk. 2, ch. 25, §§ 8–9 (1: 385).

77. Ibid., ch. 6, §§ 32–33 (1: 176–77).

78. On Solórzano, see Muldoon; Malagón and Ots Capdequí.

79. Cobo, *History*, 17.

80. Calancha, 35–39. At the turn of the seventeenth century, most learned colonists were of the opinion that the skin color of blacks and Amerindians and their purported servile behavior were not caused by environmental influences at all but originated in Noah's curse on Ham's descendants. Besides the already reviewed cases of Salinas de Córdoba and León Pinelo, see Solórzano Pereira, bk. 1, ch. 5, § 35 (1: 59); and Torquemada (1615), bk. 1, ch. 10, 33; bk. 14, ch. 15, 602–3; bk. 144, ch. 18–19, 609–14. Torquemada, who attributed the color of blacks and Amerindians to "bad blood engendered [by Noah's curse] in those poisoned bodies," also, however, described the bodies of Amerindians favorably: they enjoyed a temperate constitution, had proportionate bodies, their physiognomies revealed sharp souls, and their understandings and external senses were privileged, he said (ch. 24, 620–21); he even maintained that the heavenly constellations and climate of the Indies not only caused the land to be temperate but also its peoples to be beautiful and intelligent (ch. 25, 623). The color of the Amerindians was perceived as a singularity of nature, a puzzle to exercise the minds of philosophers, because the climate of America was varied, yet the color of Amerindians and the quality of their hair were thought to be uniform; see López de Gómara, "Del color de los Indios," fol. cxvii; Cobo, *History*, 11, 14–15. Cobo considered the color of Amerindians to be one of the greatest puzzles of America; he also considered the fact that Amerindians and Spaniards inhabited the same climatic zone yet remained two physically distinct nations to be a secret of nature that was "great indeed." On Noah's curse and European colonialism and racism, see Evans; Braude.

81. Aristotle *Politics* 7.7.269.

82. Calancha, 68. Calancha introduces his racialist paradigm in other passages, e.g., on p. 64 he argues that the fact that Amerindians, unlike Spaniards and blacks, do not suffer from goiter, mental illnesses, kidney stones, heart diseases, or asthma is attributable to a distinct Amerindian phlegmatic constitution and unrelated to what the Amerindians drink or eat (as some liked to believe).

83. Cárdenas, 92–97; 178–81; 255–60; 260–65.

84. Ibid., 203, 206–7; Creole idleness and luxury also compounded the problem, for the humors of food were not burnt off by exercise.

85. Ibid., 214–15.

86. Ibid., 241.

87. Ibid., 239.

88. Ibid., 247–49.

89. Ibid., 215.

90. The English words are the same in Spanish.

91. Barnes; Edel.

92. Cárdenas, 216.

93. Ibid., 217.

94. Ibid., 218–19.

95. Ibid., 221.

96. Ibid., 249–51.

97. Ibid., 208–13. Yet the more humid Creoles lacked constancy and perseverance (213).

98. Martínez, *Repertorio*, bk. 3, ch. 1, 157–60. According to Martínez, Mexico was also under the dominant influences of Pegasus, Taurus, and Leo, and the planet Venus; see bk. 3, ch. 3, 163–64. In bk. 1, chs. 3–6 (pp. 4–14), Martínez discusses how the planets and constellations control sublunar phenomena, including meteorological changes and human constitutions. On Martínez, see Maza, *Enrico Martínez*; Hoberman.

99. Martínez, *Repertorio*, bk. 3, ch. 2, 160–63.

100. Ibid., bk. 3, ch. 12, 178–81.

101. Ibid., bk. 3, ch. 13, 181–83.

102. Ibid., bk. 3, ch. 23, 176 (really 203; this edition contains numerous errors of pagination).

103. Ibid., bk. 3, ch. 23, 176: "que las causas universales son variadas y determinadas según la calidad de la materia, haziendo en diversos sujetos differentes effectos: el fuego consume leña seca y también la verde, mas no también y facilmente esta como aquella; muy diferente es la complexión del moreno, y también la del indio ala del Español, por lo qual las causas generales que eneste Reyno ocurren, no pueden produzir iguales effectos en todos, sino en cada uno según su temperamento, disposición del celebro y órganos corporales y desto procede la diversidad de ingenios que se halla en las referidas naciones."

104. Ibid., bk. 3, ch. 23, 177.

105. Cisneros, ch. 16, fols. 90r–103v.

106. Although Cisneros rejected the possibility of ever finding Mexico's dominant planetary influences, he did think that Mexico was under some kind of astral influences and sought to determine what they were. According to Cisneros, Taurus caused heat, dryness, and encouraged wind formations over Mexico (ch. 16, fol. 87r–v); and the land was also under the influences of three "vertical" stars (ch. 16, fols. 88v–89v). The debate over the dominant "stars and signs" of Mexico continued throughout the seventeenth century; see, e.g., Vetancurt, ch. 1, 4–5.

107. Cisneros, ch. 17, fols. 110, 112v–112v (folio 112 is paginated twice). Cisneros compares Mexican Amerindians to the description given by Hippocrates (*Of Airs, Waters, and Places*, 15) of the Colchians who lived in the region of the river Phasis, who were tall, fat, yellow, and slothful (110r–111r).

108. Cisneros, ch. 17, fols. 112v–113v.

109. Vetancurt, treatise 1, ch. 6, 12.

110. Quoted in Lockhart, 385; trans. slightly modified. In fact, Chimalpahin read Martínez's *Repertorio* and drew heavily from it. It is very likely that Chimalpahin's views on race were influenced by Martínez's. On Chimalpahin's and Martínez's relationship, see Schroeder, 82, 84; and see also ibid., 86, on Chimalpahin's negative views of mestizos.

111. Banton; Gould; Stepan, *Idea of Race*; Hudson; Hannaford; Outram, 74–79, 94–95. Alden T. Vaughan has argued that it was only after the mid eighteenth century that British American colonists began to identify themselves as whites and Amerindians as "red," whereas before Amerindians were thought to be "white." Vaughn's findings, though centered on a colonial setting, reinforce the thesis that the rise of racialist concerns was a late eighteenth-century phenomenon.

112. Hanke, *Aristotle*; Pagden, *Fall*; Seed, "'Are These Not Also Men?'"

113. Pagden, *Fall*, passim; Hulme, "Tales of Distinction," 157–97.

114. Amiel; Sicroff; Nirenberg.

115. Peña Montenegro, bk. 2, treatise 4, § 1, 176–77; see also Cabello Valboa, pt. 1, ch. 13. Cabello Valboa explains the origins of idolatry as a "contagion" that first affected the imagination of those descendants of Noah negatively influenced by some constellations, which was later transformed into an innate biological inclination (*naturaleza*) (p. 80).

116. Pagden, *Fall*, 97–104. On culture as a second nature, see also Kelley.

117. MacCormack, 383–405.

5. Eighteenth-Century Spanish Political Economy

1. To what degree most eighteenth-century Spanish scientific policies were shaped by patriotic agendas is something that merits further research. See, e.g., Clément, "De los nombres," and Pino and Guirao. On how patriotism shaped most other areas of intellectual production, see Cañizares-Esguerra, *How to Write the History of the New World*, ch. 3. On Linnaeus's views of Spain and the ensuing debate, see Pascual. On the views of Feuillée, see his "Épître" in *Journal* (1714). For a succinct and illuminating study of the development and failure of Spain's investment in botanical research, see Puerto Sarmiento, *Ilusión quebrada*.

2. Hillgarth. On the Spanish "Black Legend," see Juderías; García Carcel; Gibson, ed.; Maltby; Schmidt.

3. On travelers to Spain, see Morel-Fatio; Sarrailh, "Voyageurs français"; Aguilar Piñal; Díez Borque.

4. Elliott, *Spain and Its World*; Perdices Blas. The *arbitristas* were a variegated group, most holding very dissimilar, even opposing, ideas to one another. I have thus hedged my assertions about *arbitrismo* with qualifiers such as "many" and "some."

5. Vilar.

6. Pocock, *Machiavellian Moment*.

7. Pocock, *Virtue*; Horne; Olson, chs. 9 and 11; Hont and Ignatieff; Hirshman; Rothschild.

8. For other important characterizations of the Spanish Enlightenment, see Herr; Sarrailh, *España ilustrada*; Sánchez-Blanco Parody, *Europa y el pensamiento*; Castro; Sellés, Peset, and Lafuente; Mestre.

9. Muñoz, 8: "de sin tener conocimiento de la literatura española, sin haber quizás leído más que algún libro despreciado por nosotros mismos, nos imponen a todos la nota de ignorantes y corrompedores del buen gusto."

10. José Nicolás de Azara, letter from Rome, June 6, 1782, included in his prologue to Guillermo Bowles's *Introducción a la historia natural y a la geografía física de España*, 2d ed. (Madrid: Imprenta Real, 1782), quoted in Sarrailh, *España ilustrada*, 322n134: "Swinburne debe ser felicitado por su perpicaz penetración que, a los dos o trés días de haber entrado . . . ya había descubierto que todos los caminos eran malos, las posadas peores, el país parecido al infierno, donde reina la estupidez."

11. Forner, *Oración apologética*, 10–11: "hombres que apenas han saludado

nuestros anales, que jamás han visto uno de nuestros libros, que ignoran el estado de nuestras escuelas, que carecen del conocimiento de nuestro idioma . . . [que] en vez de acudir a tomar en las Fuentes la instrucción debida . . . echan mano, por más cómoda, de la ficción y texen a costa de la triste Península novelas y fábulas tan absurdas como pudieran nuestros antiguos Escritores de caballerías."

12. The views of long-term residents and foreign scholars who spent their lives studying Spain were welcome. See, e.g., the cases of the Irish Richard Ward and William Bowles and the French-born Count Francisco de Cabarrús, whose writings were well received and even acclaimed. The Scottish historian William Robertson, who never set foot in Spain, was made honorary member of the Spanish Royal Academy of History, although eventually his views proved controversial. On Ward, Bowles, and Cabarrús, see Sarrailh, *España ilustrada*, 323–28. On Robertson, see Cañizares-Esguerra, *How to Write the History of the New World*, 171–74.

13. A characterization of Gazel as a traveler can be found in Cadalso, *Cartas marruecas*, letters 1 and 2.

14. Cadalso, *Defensa*, 4: "sin el menor conocimiento de su historia, religión, leyes, costumbres y naturaleza."

15. Ibid., 10: "de la decadencia total de las ciencias, artes, milicias, comercio, agricultura y población."

16. Some authors did not react angrily to the views of Montesquieu, who after all used his *Lettres persanes* to ridicule French mores rather than Spanish ones. These authors agreed wholeheartedly with Montesquieu and suggested that positions like Cadalso's avoided coming to grips with the real problems of the nation. See, e.g., Capmany, 130–52.

17. Torres Villaroel, *Visiones y visitas*. A similar sensibility is to be found in *Sainetes*, a light comedy by Ramón de la Cruz (1731–94).

18. Cadalso, *Eruditos a la violeta*, 167.

19. Juan de Iriarte quoted in Cotarelo y Mori, 16: "Componer apologías de la lengua patria en vindicación de las calumnias extranjeras, que hasta nacionalidad le niegan, haciéndole africana o asiática algunos, circunscribiendo otros a uno o dos el número de nuestros buenos autores, y afirmando que toda la ciencia de España se reduce a dos coplas y cuatro silogismos. . . . [El rol de la Academia consiste] en hacer elogio de los grandes y esclarecidos varones de nuestra patria resucitando sus memorias y sus nombres."

20. Mestre; François Lopez, *Juan Pablo Forner*.

21. François Lopez, "Comment l'Espagne eclairée inventa le Siècle d'Or"; Juárez Medina.

22. Gándara, 53–81.

23. Ibid., 86.

24. Arroyal, 9: "Mientras no se descubra [la causa de la decadencia española] será vano cualquier esfuerzo que se haga para contener sus funestos efectos."

25. On Jovellanos, see Varela.

26. Jovellanos, 95: "con gran detenimiento y circunspección para no aventurar el descubrimiento de la verdad en una materia en que los errores son de tan general y perniciosa influencia."

27. Ibid., 104: "el problema no tanto estriba en presentarle estímulos [a la agricultura] como en remover los estorbos que retardan su progreso."

28. Sempere y Guarinos, *Historia del luxo*, 1: 8: "qual es el de formarse generalmente ideas falsas, e inexactas, acerca de los más importantes puntos de la legislación, y la política; confundirse frequentemente las causas con los efectos; atribuirse a unas los que lo son de otras muy diferentes: de donde proviene el promulgarse leyes, no solamente inútiles, sino muchas veces contrarias al objeto, y a las intenciones de los mismos legisladores que las expidieron."

29. Ibid., 2: 176–218.

30. Sempere y Guarinos, *Consideraciones*, 158 (quotation), 79; 118–23, 157–58, 179–80 (for examples of elite ignorance of political economy).

31. Herr, 219–30 and passim. This interpretation of the Morvilliers affair actually first originated in the mid nineteenth century with Marcelino Menéndez y Pelayo, who denounced contemporary Spanish liberals as shallow Franco-Anglophiles, heirs to those who had taken Morvilliers's side in the debate. Forner, on the other hand, appeared as an authentic Spanish intellectual. Emilio Cotarelo y Mori, a leading fin de siècle scholar, simply inverted the heroes in Menéndez y Pelayo's account, helping to perpetuate this Manichean narrative. See Marcelino Menéndez y Pelayo, "Mr. Masson, redivivo," *Revista Europea* 8, 127 (July 30, 1876), and "Mr. Masson, redimuerto," ibid., no. 135 (September 24, 1876), reproduced in García Camarero and García Camarero, 209–30, 239–68; Cotarelo y Mori, 312–22.

32. Maravall, "Sentimiento de nación"; François Lopez, *Juan Pablo Forner*. On Forner, see also see Sánchez-Blanco Parody, *Absolutismo y las luces*.

33. Sempere y Guarinos, *Ensayo de una biblioteca*.

34. Masdeu; Lampillas; Andrés.

35. Forner, *Oración apologética*, passim.

36. Forner, *Discurso sobre el modo de escribir*, 165: "al modo de los que forman sistemas [muchos] echaron el peso que ocasionó la ruina sobre un solo defecto, y de él fueron derivando la serie de males que se atropellan después para enflaquecer y debilitar la monarquía."

37. Ibid., 166: "buenas y malas, útiles y perniciosas, sabias y desconcertadas."

38. It is not entirely clear whether the editor was Luis Cañuelo. See José Miguel Caso González, "Estudio," in *El Censor: Obra Periódica*. On *El Censor* and the project it represented, see Sánchez-Blanco Parody, *Absolutismo y las luces*.

39. *El Censor*, discursos 84–85 (December 22 and 29, 1785).

40. Ibid., discurso 54 (January 1, 1784); discursos 124–27 (September 28 to October 19, 1786).

41. Ibid., discurso 22 (July 5, 1781); discurso 52 (December 18, 1783).

42. Ibid., discurso 53 (December 25, 1783); discurso 65 (March 18, 1784).

43. Ibid., discurso 59 (February 5, 1784); discurso 79 (November 17, 1785); discurso 81 (December 1, 1785).

44. Ibid., discurso 120 (August 31, 1786); discurso 165 (August 5, 1787).

45. See Hobbes. On bodily metaphors of the polity in Spain, see Rico. For studies of the metaphor in general, see Barkan; Spary, "Political, Natural, and Bodily Economies."

46. *El Censor*, discurso 157 (June 14, 1787).

6. How Derivative Was Humboldt?

1. Botting; Kellner. On Humboldt's new science and methodologies, see Cannon; Dettelbach, "Humboldtian Science."

2. Browne; Nicolson; Bowen; Dettelbach, "Global Physics."

3. Brading, *First America*, 526–32. For an important study of Humboldt that takes Spanish America Creole society seriously, see Minguet.

4. See Cañizares-Esguerra, *How to Write the History of the New World*.

5. On Caldas, see Appel. On Mutis, see Frías Núñez, *Tras el Dorado vegetal*. On the practice of natural history in late eighteenth-century Spanish America, see Nieto Olarte. On the almost complete annihilation of a generation of patriot naturalists in the Spanish American colonies, particularly in New Granada, see Glick.

6. All these unpublished maps and memoirs by Caldas are kept at the Archivo del Real Jardín Botánico de Madrid, División III, Serie Botánica, Fondo Mutis. The full title of Humboldt's essay is *Essai sur la géographie des plantes: Accompagné d'un tableau physique des régions équinoxiales, fondé sur des mesures exécutées, depuis le dixième degré de latitude boréale jusqu'au dixième degré de latitude australe, pendant les années 1799, 1800, 1801, 1802 et 1803* (Paris: Levrault, Schoell, 1805).

7. Francisco José de Caldas, letters to Santiago Arroyo, July 20, September 21, and October 6, 1801, in Chenu, 107, 131, 133. The original of the first quoted passage reads: "podemos esperar algo útil y sabio de un hombre que va a atravesar el Reino con la mayor velocidad? . . . Quién sabe si va a llenar de preocupaciones y de falsas noticias a la Europa, como lo han hecho casi todos los viajeros."

8. For a recent example of this type of patriotic literature, see Arias de Greif.

9. Linnaeus, "On the Increase of the Habitable Earth."

10. Columbus. On paradise in the supralunar sphere, see Singleton.

11. José de Acosta, *Historia natural y moral*, 1987 ed., bk. 2, passim, esp. chs. 12, 14; bk. 3, chs. 19–20.

12. These are some of the early modern authors León Pinelo set out to engage: Jan Becan, *Origines Antwerpianae* (Antwerp, 1569); Guillaume Postel, *Cosmographie disciplinae compendium* (Basel, 1561); Franciscus Lucas Brugensis et al., eds., *Biblia Sacra* (Louvain, 1581); Jacques d'Auzoles, *La sainte géographie, c'est à dire, exacte description de la terre et véritable démonstration du paradis terrestre* (Paris, 1629). On early modern literature on the historical location of paradise, see Delumeau, ch. 8.

13. On marvels and curiosities in the early modern period, see Daston and Park.

14. For his views of the Andes as a privileged space, see León Pinelo, *Paraíso*, 1: 307–13, 383–96.

15. On the thoroughly multinational and multiethnic character of the loosely held Spanish Catholic monarchy on the eve of the Bourbon reforms, see Kamen. For an illuminating continentwide comparative perspective on the social and political changes that led to the wars of independence in the Americas, see Rodríguez O., "Emancipation of America." Rodríguez takes issue with traditional accounts that see British America undergoing a radical political "revolution" and Spanish America experiencing none. For the most recent statement of this tiresome narrative, see Langley.

16. Lafuente and Mazuecos; Steele; González Bueno, ed.; Puerto Sarmiento, *Ilusión quebrada*; Lozoya; Engstrand; Pimentel, *Física de la monarquía*; Frías Núñez, *Tras el dorado vegetal*.

17. José de Gálvez quoted in Puerto Sarmiento, *Ciencia de cámara*, 155–56: "doce naturalistas con otros tantos chymicos o mineralogistas esparcidos por sus estados, producirán por medios de sus peregrinaciones una utilidad incomparablemente mayor, que cien mil hombres combatiendo por añadir al Imperio Español algunas provincias."

18. On this expedition, see Frías Núñez, *Tras el Dorado vegetal*.

19. See Mutis's article on quinine in *Diario de Madrid*, no. 315 (November 11, 1880), reproduced in *Flora de la Real Expedición Botánica del Nuevo Reino de Granada*, 44: 42–43.

20. Mutis, *Escritos científicos* 1: 177: "era como un centro de las Américas en el que artículos similares o equivalentes a aquellos que pueden encontrarse en el inmenso espacio del Viejo y el Nuevo mundo se han reunido." On this project, see Frías Núñez, "Té de Bogotá."

21. Vargas, 6: "las diversas alturas de este Reino sobre la superficie del mar, y sus diversas distancias a la línea, hacen que en su Distrito se hallen casi todos los temperamentos del globo, y en algunas partes tan inmediatos unos a otros que en un día se puede experimentar frío por la mañana, temperamento medio u otoño al mediodía y excesivo calor por la noche, según se baja de las cordilleras."

22. Zea, 68.

23. Sabio patriota, "Al señor autor del periódico," *Papel Periódico*, no. 11 (April 22, 1791): 81. Many other articles emphasized the same theme of New Granada as microcosm of the world; e.g., see Observador amigo del país, "Discurso," ibid., 86; Luis de Atigarraga, "Disertación sobre la agricultura dirigida a los habitantes del Nuevo Reyno de Granada," ibid., no. 56 (March 9, 1792): 36–37; Diego Martín Tanco, "Discurso por el cual se manifiestan los medios de aumentar la población de este reyno," ibid., no. 76 (July 27, 1792), 197; "Idea del nuevo Reyno de Granada," ibid., no. 256 (August 12, 1796), 1537–38.

24. Caldas, "Ensayo sobre el estado de la geografía," 276: "mejor situada que Tiro y que Alexandría, [Nueva Granada] puede acumular en su seno los perfumes del Asuam, el marfil africano, la industria europea, las pieles del Norte, la ballena del Mediodía y cuanto produce la superficie de nuestro globo."

25. Caldas, "Influjo del clima," 112: "hay pocos puntos sobre la superficie del globo más ventajosos para observar, y puedo decir, para tocar el influjo del clima y de los alimentos sobre la constitución física del hombre, sobre su carácter, sus virtudes y sus vicios." See also Caldas, "Ensayo sobre el estado," 275.

26. Caldas, "Ensayo sobre el estado," 276–77: "La posición geográfica de la Nueva Granada parece que la destina al comercio del universo."

27. Tadeo Lozano quoted in Hernández de Alba, 148: "en unos cuantos siglos en un vasto imperio que . . . igualará al más poderoso de Europa."

28. Unanue, "Geografía física," 11: "Parece que después de haberse exercitado [Dios] en los abrasados arenales del Africa, en los frondosos y fragantes bosques del Asia, en los climas templados y fríos de la Europa, se esfuerza a reunir en el Perú quantas producciones había esparcido en aquellas tres partes, para formarse

un templo digno de su inmensidad, y reposar en el magestuosamente cercada de todas ellas; tantas son las riquezas que encierra este admirable Reyno." See also ibid., 16: Peru as a temple of God has its façade to the north: its dome is the celestial vault at the equator; its columns are the mountains; and its perpetual lamps are the volcanoes.

29. Ibid., 21. Like Caldas, Unanue thought that the Andes was a privileged laboratory to study the influence of climate on humans; see Unanue, *Observaciones*, 47, 171.

30. Unanue, "Geografía física," 22–26: "El Perú es la obra de más magnificencia que ha criado la Naturaleza sobre la tierra." For a detailed analysis of Unanue's views, see Cañizares-Esguerra, "Utopía de Hipólito Unanue."

31. Unanue, "Disertación sobre el aspecto . . . Coca," 241–45.

32. Cervino, "El tridente de neptuno es el cetro del mundo: Discurso inaugural de la Academia Naútica, del 25 noviembre de 1799," reproduced in Chiaramonte, 295: "Nuestra ubicación [en el globo terráqueo] es una muy feliz [porque] América del Norte, Europa, Asia y el Océano Pacífico están equidistantes de nosotros. Esta ubicación maravillosa nos asegura un gran tráfico comercial. [Nosotros] nos convertiremos en el almacén del mundo."

33. Caballe, "Continua la idea general del comercio de las provincias del Río de la Plata," *Telégrafo Mercantil, Rural, Politico-economico, e Historiografo del Río de la Plata* 4 (April 11, 1801), reproduced in Chiaramonte, 227 and 229 (on the privileged central position of Buenos Aires): "toda suerte de resinas y fármacos, sin mencionar oro y plata preciosos y en abundancia . . . [así como los igualmente preciosos] salitre, perlas y conchas marinas que pueden ser encontrados en abundancia en el inmenso Chaco"; "Sin recurso a la hipérbole [se concluye que] alrededor del mundo no hay otra tierra tan rica, que posea tal variedad de productos . . . y [por lo tanto] tan apta para establecer instituciones comerciales fuertes y poderosas."

34. "Comercio," *Seminario de Agricultura* 4 (October 13, 1802), reproduced in Chiaramonte, 266–67: "como un mar, [en el que] nos perdemos en el horizonte . . . una tierra de montañas fabulosas con la mejor madera del universo." See also "Agricultura," *Seminario de Agricultura* 1 (September 1, 1802), reproduced in Chiaramonte, 254: "localizada [justo] en el centro del mundo comercial y deliciosamente situada a las orillas de un poderoso río"; "con el poder productivo más grande del globo."

35. San Vicente, 32, 34.

36. Venegas, advertencia, n.p.

37. Moziño, reproduced in Trabulse, *Historia de la ciencia*, 116–17.

38. Drayton; Spary, *Utopia's Garden*; Smith and Findlen; Miller and Reill; Koerner; Schiebinger and Swan.

7. LANDSCAPES AND IDENTITIES

1. Orozco y Berra, 429–33. Orozco y Berra's upbeat narrative contrasts sharply with that offered by Antonio García y Cubas, the other giant of nineteenth-century Mexican geography. García y Cubas acknowledges that despite the anarchy, local scholarship on geography advanced in the course of the century, but very slowly.

Owing to the internecine wars, the different factions have used the landscape itself to fight their battles, christening places after their heroes and thus constantly changing the names of cities, towns, and geographical sites. See García y Cubas, v–vii. On the history of nineteenth-century Mexican cartography, see Craib; Mendoza Vargas.

2. For two interesting studies in English, see Ades, 101–9; Widdifield, 64–77. Dawn Ades argues that Velasco was not a *plein-air* naturalist in the tradition that beginning with the Barbizon school evolved into Monet's impressionism. Velasco, Ades suggests, was a rather traditional academic painter who nevertheless tapped into the topographical and panoramic visual idioms of foreign traveler-artists en route through Mexico as he created original compositions. Stacie G. Widdifield, on the other hand, deals with Velasco as part of a much larger and ambitious study of the ideological functions of nineteenth-century Mexican academic painting. Widdifield concludes that landscapes became popular only after the various nineteenth-century civil wars and military foreign occupations ended and prosperity and stability ensued. Velasco's landscapes sought to prove to foreign audiences that Mexico had finally overcome its checkered past and embarked on a path of enlightened cosmopolitanism and progress

3. On Velasco, see Moyssén Echeverría; Moyssén Echeverria et al.; Altamirano Piolle; Trabulse, *José María Velasco*; Encina; Pérez de Salazar y Solana. On other landscape painters and the genre itself, see Lagos; Caballero-Barnard; Barrios; Romero de Terreros, "Descubridores del paisaje mexicano" and *Paisajistas mexicanos*; and the catalogues *México una visión de su paisaje*; *Paisaje and otros paisajes mexicanos del siglo XIX en la Colección del Museo Soumaya*; and *Joaquín Clausell y los ecos del impresionismo en México*.

4. On Brazil, see the works of Carlos Roberto Maciel Levy, who single-handedly created the field: *O Grupo Grim, paisagismo brasileiro no século XIX* (Rio de Janeiro: Edições Pinakotheke, 1980); *Giovanni Battista Castagneto (1851–1900) o pintor do mar* (Rio de Janeiro: Edições Pinakotheke, 1982); *Antônio Parreiras (1860–1937), pintor de paisagem: gênero e história* (Rio de Janeiro: Edições Pinakotheke, 1981); *150 Anos de pintura de marinha na história da arte brasileira* (Rio de Janeiro: Museu Nacional de Belas Artes, 1982). See also Nagib Francisco, *João Batista da Costa, 1865–1926* (Rio de Janeiro: Edições Pinakotheke, 1984), and José Maria dos Reis Júnior, *Belmiro de Almeida, 1858–1935* (Rio de Janeiro: Edições Pinakotheke, 1984). On Colombia, see Eduardo Serrano, *La Escuela de la Sabana* (Bogotá: Museo de Arte Moderno de Bogotá, 1990); Juan Luis Mejía Arango, ed., *Poesía de la naturaleza* (Medellín: Compañia Sudamericana de Seguros, 1997). On Ecuador, see Alexandra Kennedy Troya, *Rafael Troya: El pintor de los Andes ecuatorianos* (Quito: Banco Central del Ecuador, 1999).

5. Ann Jensen Adams, "Competing Communities in the 'Great Bog of Europe': Identity and Seventeenth-Century Dutch Landscape Painting," in W. J. T. Mitchell, ed., *Landscape and Power* (Chicago: University of Chicago Press, 1994), 35–76; Mariët Westermann, *A Worldly Art: The Dutch Republic, 1585–1718* (New York: Perspectives/Harry N. Abrams, 1996).

6. D. B. Brown, 140–50; J. L. Koerner.

7. Bonyhady; Astbury.

8. Kelly. On the misleading conflation that has a regional landscape (New York) stand for the nation, and on how that took place in the nineteenth century, see A. Miller.

9. *Tres grandes maestros del paisaje decimonónico español*. On Spanish landscape painting, see also Litvak.

10. Ortega Cantero.

11. Pena, *Pintura de paisaje e ideología*; Jurkevich.

12. *L'Escola d'Olot*.

13. Adams.

14. Reyes; Millán.

15. Roca, 25.

16. For one of the earliest published references in colonial sources to the sacred status of *ahuehuetes*, see Valadés, bk. 3, ch. 3, 168, who says the natives planted them in gardens near Mesoamerican pyramidal temples. Since they were huge, one tree alone could provide shade to thousands of natives. Despite not bearing fruit, these trees were so precious that they were used to assess the worth of all other trees. On account of this Spaniards called the *ahuehuete* the "tree of paradise": "Semper autem in illis plantabant magno studio arbores valde patulas & umbrosas, usque adeo, ut in unius umbra mille hominess agree possint, eo modo, quo Indi sedent. Quamuis autem sterilis & infrugifera sit ea arbor, est nihilominus in tanto pretio, ut in comparationibus maiori plerunque ab ipsa collationem ducant. Vocant autem illam Indi ahuehuetl, Hispani arbor de parayso." Valades also includes *ahuehuetes* right behind a temple in his illustration of ritual sacrifices by natives in the New World, particularly in Mexico: "Tipus sacrificiorum que in manitis Indi faciebant in Novo Indiarum Orbe, praecipue in Mexico" (p. 177). Agustín Dávila Padilla in his (1596) history of the Dominicans also makes reference to *ahuehuetes* growing in courtyards in front of every temple (see Dávila Padilla, bk. 1, ch. 24, 75). Francisco Hernández, who compiled a vast natural history of Mexico from 1571 to 1577, describes four varieties of the tree but makes no reference to their sacred status, simply to their medical virtues and their tendency to grow in lacustrine areas. See Nieremberg, *Historia natvrae*, bk. 14, ch. 27, 305–6; Hernández, *Rerum medicarum Novae Hispaniae thesaurus*, bk. 3, ch. 66, 93–94.

17. See, e.g., Noriega; Ortega Reyes. Ortega Reyes uses the extraordinarily big *ahuehuete* at the town of Santa María de Tule, Oaxaca, not only to praise as providential the bountifulness of the land but also to link the tree to the history of the Zapotec civilization of Mitla.

18. On Altamirano's sensibilities toward the landscape, see Giron; Díaz y de Ovando.

19. Payno; Altamirano, *Paisajes y leyendas*.

20. Altamirano, *Paisajes y leyendas*, 8.

21. On an indigenous tradition that uses toponyms as shorthand to narrate complex historical memories, see also Carson.

22. Cañizares-Esguerra, *How to Write the History of the New World*.

23. There was another myth on the origin of the volcanoes also cast as a narrative of *mestizaje*: Ixtaccihuatl, a beautiful maiden whose hair was as black as ebony, whose lips were as red as fire, and whose skin was as *white* as snow, was the daughter

of Sky Lord and Mother Earth. After losing his first wife, father Sky Lord married again. The jealous stepmother, however, enraged by Ixtaccihuatl's beauty, poisoned the girl, who fell into a comatose state. Popocateptl, an Amerindian prince, discovered the sleeping beauty and caused an eruption to wake her up. The couple married and lived happily ever after.

24. Altamirano, "Salón en 1879–1880."

25. Hamblyn.

26. This paragraph and those to come that deal with Velasco's biography are based on the many works on Velasco cited in note 3 above, especially on the biography by Altamirano Piolle.

27. On the exquisite knowledge of the various branches of natural history packed into most landscape paintings of the Hudson River School, see Novak; Bedell. On Velasco and his science, see Trabulse, *José María Velasco*.

28. See, e.g., Velasco, "Estudios sobre la familia de las cacteas de México" and "Estudio sobre una nueva especie de falsa jalapa de Querétaro." Velasco completed botanical illustrations for the ill-fated periodical *Flora botánica de México* (1868) and his friend's Fernando Altamirano Carvajal's *Leguminosas medicinales autóctonas* (1878), a medical herbal that sought to build on Francisco Hernández's tradition of indigenous knowledge. Velasco also collaborated with illustrations in two studies of hummingbirds: Manuel Villada's *Troquilideos del Valle de México* (1873) and Rafael Montes de Oca's *Ensayo ornitológico de la familia Troquilidae o sea de los colibríes o chupamirtos de México* (1876).

29. Velasco, "Descripción, metamórfosis y costumbres" and "Estudio anatómico."

30. Velasco, "Anotaciones y observaciones al trabajo del señor Augusto Weismann." Casting himself as a privileged observer is redolent of earlier colonial forms of patriotic epistemologies. On the latter, see Cañizares-Esguerra.

31. On Velasco's role as one of the leading figures in the official efforts to have Mexico participate in World Fairs in the second half of the nineteenth century, see Tenorio-Trillo, chs. 4 and 7.

32. Manthorne; Kelly; Miller.

33. Guadalajara, Jalisco, was the only city in Mexico in which artists such as Francisco and Jacobo Gálvez and Gerardo Suárez painted scenes of cattle ranching. On Jalisco's school of painting, see Camacho. On the works of Francisco Gálvez, see *Jalisco Genio y Maestría*.

34. See Wobeser.

35. On the urban bias of Mesoamerican indigenous identities, see Farriss, *Maya Society Under Colonial Rule*, and Lockhart, *The Nahuas After the Conquest*.

36. Pratt, pt. 2.

37. Manthorne.

38. Brading, *Origins of Mexican Nationalism*.

39. Altamirano, "Salón en 1879–1880."

40. The Mexican Society of Geography and Statistics decided to break a meteorite that fell in San Luis Potosí into pieces to study its mineralogical composition. The Society of Natural History vehemently opposed such destruction. The episode sparked a bitter debate between these institutions, which took on political overtones. See "Dictamen aprobado por la Sociedad de Historia Natural, en la sesión

del 17 de Abril de 1873, . . . para dilucidar la cuestión suscitada con motivo del fraccionamiento del aerólito de la 'Descubridora,'" *La Naturaleza*, 1st ser., 2 (1871–73): 277–97, esp. 293.

41. Encina is particularly good at highlighting Velasco's conservative Catholic temperament.

42. Ramírez. The fact that Velasco was painting historically charged landscapes to celebrate figures like Juárez and Porfirio Díaz in the 1880s shows that Mexican conservatives saw the latter's rule as a good substitute for their long-sought-after Catholic constitutional monarchy. On the Porfiriato as an assault on the liberal doctrine of innate human rights under the constitution, see Hale.

43. Heredia, 54.

44. Alcaraz, 95–123.

45. On the geological views represented in the paintings of Cole, Church, Haseltine, and other nineteenth-century U.S. artists, see Bedell.

46. Vulcanism survived throughout the nineteenth century. In addition to Heredia and Alcaraz, see, e.g., Carpio; Romero; Othón, "Las montañas épicas" (1902), in *Paisaje*.

47. Carpio, 135–41.

48. For a recent new synthesis, see Brading, *Mexican Phoenix*.

49. For examples of this genre, see Pesado; Elguero; Carpio, "A la Virgen de Guadalupe" (1849), in *Poesías*, 93–97. For the colonial origins of the trope of ecological transformation, see Cañizares-Esguerra, *Puritan Conquistadors*, ch. 5.

50. Tenorio-Trillo, chs. 5 and 9. This seems to have been a widely held view of racial *mestizaje* throughout Latin America. See, e.g., Stephan; Graham.

51. Widdifield, ch. 4

52. *Joaquin Clausell y los ecos del impresionismo en México.*

53. Rafael Angel de la Peña, "Prólogo," in Pegaza, vi.

54. Othón, "Pastoral" (*Poemas Rústicos*, 1902), in *Paisaje*, 41.

55. On Othón's views on written descriptions of landscape as painting, see his "Ocaso" (*Poemas rústicos*, 1902), in *Paisaje*, 11.

56. See, e.g., Othón's "Himno de los Bosques" (*Poemas rústicos*, 1902), in *Paisaje*, 85–97.

57. Johns, 6.

58. Gay, 4.

Select Bibliography

ARCHIVES

Archivo de la Real Academia de Historia, Madrid
Archivo del Real Jardín Botánico, Madrid
Archivo General de Indias, Seville, Indiferente General
Biblioteca Nacional de México, Fondos Reservados (manuscritos)

BOOKS

Acosta, Cristóbal. *Tractado de las drogas, y medicinas de las Indias Orientales. . . .* Burgos: M. de Victoria. 1578.
Acosta, José de. *De Christo revelato libri novem. Simulque De Temporibus novissimis libri quator.* Lyon: Ioannem Baptistam Buysson, 1592.
———. *Historia natural y moral de las indias en que se tratan cosas notables del cielo y elementos, metales, plantas y animales dellas.* 1590. Reprint. Barcelona: Iayme Cendrat, 1591.
———. *Historia natural y moral de las Indias.* Edited by José Alcina Franch. Madrid: Historia 16, 1987.
———. *Natural and Moral History of the Indies.* 1590. Edited by Jane E. Mangan, with an introduction and commentary by Walter Mignolo. Translated by Frances López-Morillas. Durham, N.C.: Duke University Press, 2002.
Adams, Steven. *The Barbizon School and the Origins of Impressionism.* London: Phaidon, 1994.
Ades, Dawn. "José María Velasco." In Dawn Ades, ed., *Art in Latin America: The Modern Era, 1820–1980,* 101–9. New Haven, Conn.: Yale University Press, 1989.
Aguilar Piñal, Francisco. "Relatos de viajes de extranjeros por la España del siglo XVIII: Estudios realizados hasta el presente." *Boletín del Centro de Estudios del Siglo XVIII,* 4–5 (1977): 203–8.
Alatas, Syed Hussein. *The Myth of the Lazy Native: A Study of the Image of the Malays, Filipinos, and Javanese from the 16th to the 20th Century and Its Function in the Ideology of Colonial Capitalism.* London: F. Cass, 1977.
Albuquerque, Luis de. *Historia de la navegación portuguesa.* Madrid: MAPFRE, 1991.

Alcaraz, Ramón. *Poesías*. Vol. 2. Mexico City: Imprenta de Ignacio Cumplido, 1860.

Alexander, Amir R. *Geometrical Landscapes: The Voyages of Discovery and the Transformation of Mathematical Practice*. Stanford, Calif.: Stanford University Press, 2002.

Allen, Don Cameron. *The Star-Crossed Renaissance: The Quarrel About Astrology and Its Influences in England*. Durham, N.C.: Duke University Press, 1944.

Altamirano, Ignacio. "El Salón en 1879–1880: Impresiones de un aficionado." *La Libertad*, January 18, 1880. Reproduced in Ida Rodríguez Prampolini, *La crítica de arte en México* (Mexico City: Universidad Nacional Autónoma de México, 1997), 3: 32–33.

———. *Paisajes y leyendas: Tradiciones y costumbres de México*. Parts 1 and 2. Mexico City: Porrúa, 1999.

Altamirano Piolle, María Elena, ed. *National Homage: José María Velasco (1840–1912)*. 2 vols. Mexico City: Amigos del Museo Nacional del Arte, 1993.

Alvarez Peláez, Raquel. *La conquista de la naturaleza Americana*. Madrid: Consejo Superior de Investigaciones Científicas, 1993.

Amiel, Charles. "La 'pureté de sang' en Espagne." *Etudes inter-ethniques* 6 (1983): 28–45.

Andrés, Juan. *Origen, progresos y estado actual de toda la literatura*. 10 vols. Madrid: Sancha, 1784–1806.

Anghiera, Pietro Martire d'. *De orbe novo*. Alcala de Henares: Arnaldi Guillelmi, 1516.

———. *The decades of the newe worlde or west India . . . Written in the Latine tounge by Peter Martyr of Angleria, and translated into Englysshe by Rycharde Eden*. London, 1555.

Añón Feliú, Carmen. "La Granja: Castilian Baroque and European Classicism." In Monique Mosser and Georges Teyssot, eds., *The Architecture of Western Gardens: A Design History from the Renaissance to the Present Day*, 198–201. Cambridge, Mass.: MIT Press, 1991.

———. "L'immagine della natura nell'Escorial di Filippo II." *Resauro-Cittá* 5–6 (1987).

———. "Nature and the Idea of Gardening in Eighteenth-Century Spain." In Monique Mosser and Georges Teyssot, eds., *The Architecture of Western Gardens: A Design History from the Renaissance to the Present Day*, 281–92. Cambridge, Mass.: MIT Press, 1991.

Añón Feliú, Carmen, and José Luis Sancho, eds. *Jardín y naturaleza en el reinado de Felipe II*. Madrid: Sociedad Estatal para la Conmemoración de los Centenarios de Felipe II y Carlos V, 1998.

Antonio, Nicolás. *Bibliotheca Hispana[nova] sive Hispanorum, qvi vsqvam vnqvamve sive latina sive populari sive alia quavis lingua scripto aliquid consignaverunt notitia*. Rome: N. A. Tinassii, 1672.

———. *Bibliotheca Hispana vetus sive Hispanorou*. Rome: Antonii de Rubeis, 1696.

Appel, John Wilton. *Francisco José de Caldas: A Scientist at Work in New Granada*. Philadelphia: American Philosophical Society, 1994.

Arber, Edward, ed. *The First Three English Books on America*. Birmingham: n.p., 1885.

Arboleda, Luis Carlos. "Acerca del problema de la difusión científica en la periferia: El caso de la física newtoniana en la Nueva Granada." *Quipu (Revista Latinoamericana de la Historia de la Ciencia y la Tecnología)* 4 (1987): 7–32.

Arias de Greif, Jorge. "Encuentro de Humboldt con la ciencia en la España Americana: Transferencias en dos sentidos." In Frank Holl, ed., *El Regreso de Humboldt: Exposición en el Museo de la Ciudad de Quito, Junio–Agosto del 2001*, 33–41. Quito: Municipio de Quito, 2001.

Aristotle. *The Politics*. Translated by T. A. Sinclair. Baltimore: Penguin Books, 1962.

Armitage, David. *The Ideological Origins of the British Empire*. Cambridge: Cambridge University Press, 2000.

Arroyal, León de. *Cartas económico-políticas (con la segunda parte inédita)*. 1790s. Edited by José Caso González. Oviedo: Universidad de Oviedo, 1971.

Astbury, Leigh. *City Bushmen: The Heidelberg School and the Rural Mythology*. Melbourne: Oxford University Press, 1985.

Bacon, Francis. *The New Atlantis: A work unfinished*. London: William Lee, 1627.

Bakewell, Peter. "Technological Change in Potosí: The Silver Boom of the 1570s." *Jahrbuch für Geschichte von Staat, Wirtschaft und Gesellschaft Lateinamerikas* 14 (1977): 57–77.

Baldó, Marc. "La Ilustración en la Universidad de Córdoba y el Colegio de San Carlos de Buenos Aires (1767–1810)." *Estudios de historia social y económica de América* (Alcalá) 7 (1991): 31–54.

Banton, Michael. *The Idea of Race*. London: Tavistock Publications, 1977.

Bargalló, Modesto. *La minería y la metalurgia en la America española durante la época colonial*. Mexico City: Fondo de Cultura Económica, 1955.

Barkan, Leonard. *Nature's Work of Art: The Human Body as Image of the World*. New Haven, Conn.: Yale University Press, 1975.

Barnes, Jonathan. *Aristotle*. Oxford: Oxford University Press, 1982.

Baroni Vannucci, Alessandra. *Jan van der Straet detto Giovanni Stradano: Flandrus pictor et inventor*. Milan: Jandi Sapi, 1997.

Barrera, Antonio. *Experiencing Nature: The Spanish American Empire and the Early Scientific Revolution*. Austin: University of Texas Press, 2006.

———. "Local Herbs, Global Medicines: Commerce, Knowledge, and Commodities in Spanish America." In *Merchants and Marvels: Commerce, Science, and Art in Early Modern Europe*, ed. Pamela H. Smith and Paula Findlen, 163–81. New York: Routledge, 2002.

Barrios, Luisa. *Luis Coto*. Toluca: Instituto Mexiquense de Cultura, 1997.

Basso, Keith H. *Wisdom Sits in Places: Landscape and Language Among the Apaches*. Albuquerque: University of New Mexico Press, 1996.

Beaumont, Pablo de la Purísima Concepción [Juan Blas]. *Tratado de la agua mineral caliente de San Bartolomé*. Mexico City: Joseph Antonio de Hogal, 1772.

Bedell, Rebecca. *The Anatomy of Nature: Geology and American Landscape Painting, 1825–1875*. Princeton, N.J.: Princeton University Press, 2001.

Bermúdez, Joseph Manuel. "Discurso sobre la utilidad e importancia de la lengua general del Perú." *Mercurio Peruano* 9 (1793): 178–79.

Biagioli, Mario. *Galileo, Courtier: The Practice of Science in the Culture of Absolutism.* Chicago: University of Chicago Press, 1993.

Boas, Marie. *The Scientific Renaissance, 1450–1630.* New York: Harper & Row, 1962.

Bonneville, Zacharie de Pazzi de. *De l'Amerique et des Americains, ou, Observations curieuses de philosophe La Douceur, qui a parcouru cet hemisphere pendant la derniere guerre, en faisant le noble metier de tuer des hommes sans les manger.* Berlin: Samuel Pitra, 1771.

Bonyhady, Tim. *Images in Opposition: Australian Landscape Painting, 1801–1890.* Melbourne: Oxford University Press, 1985.

Born, Ignaz von, baron. *Über das Anquicken der gold- und silbehaltigen Erze, Rohsteine, Schwarzkupfer und Hüttenspeise.* Vienna: C. F. Wappler, 1786. Translated by R. E. Raspe as *Baron Inigo Born's new process. Of amalgamation of gold and silver ores, and other metallic mixtures . . .* (London: T. Cadell, 1791).

Borunda, Ignacio. "Clave general de geroglíficos Americanos." Ca. 1792. In Nicolás León, ed., *Biblioteca mexicana del siglo XVIII,* sec. 1, pt. 3. *Boletín del Instituto Bibliográfico Mexicano* 7 (1906).

Botero, Giovanni. *Le relationi vniversali di Giovanni Botero Benese. . . .* Venice: Giorgio Angelieri, 1596.

Botting, Douglass. *Humboldt and the Cosmos.* New York: Harper & Row, 1973.

Bottoni, Federico. *La evidencia de la circulación de la sangre.* 1723. Reproduced in Alvar Martínez Vidal, ed., *El nuevo sol de la medicina en la ciudad de los reyes* (Zaragoza: Comisión Aragonesa Quinto Centenario, 1990).

Bouza, Fernando. *Corre manuscrito: Una historia cultural del Siglo de Oro.* Madrid: Marcial Pons, 2001.

Bowen, Margarita. *Empiricism and Geographical Thought: From Francis Bacon to Alexander von Humboldt.* Cambridge: Cambridge University Press, 1981.

Brading, David. *First America, the Spanish Monarchy, Creole Patriots, and the Liberal State, 1492–1867.* Cambridge: Cambridge University Press, 1991.

———. *Mexican Phoenix: Our Lady of Guadalupe. Image and Process.* Cambridge: Cambridge University Press, 2001.

———. *The Origins of Mexican Nationalism.* Cambridge: University of Cambridge Centre of Latin American Studies, 1985.

Braude, Benjamin. "The Sons of Noah and the Construction of Ethnic and Geographical Identities in Medieval and Early Modern Periods." *William and Mary Quarterly,* 3d ser., 54 (1997): 103–42.

Brown, David Blayney. *Romanticism.* London: Phaidon, 2001.

Brown, Kendall. "La recepción de la tecnología minera española en las minas de Huancavelica, siglo XVIII." In *Saberes andinos: Ciencia y tecnología en Bolivia, Ecuador y Peru,* ed. Marcos Cueto, 59–90. Lima: Instituto de Estudios Peruanos, 1995.

Brown, Peter. *The Body and Society: Men, Women, and Sexual Renunciation in Early Christianity.* New York: Columbia University Press, 1988.

Brown, Thomas. *La Academia de San Carlos en Nueva España.* Mexico City: Septentas, 1976.

Browne, Janet. *The Secular Ark: Studies in the History of Biogeography.* New Haven, Conn.: Yale University Press, 1983.

Bustamante, Jesús. "De la naturaleza y los naturales americanos en el siglo XVI: Algunas cuestiones críticas sobre la obra de Francisco Hernández." *Revista de Indias* 52 (1992): 297–328.

———. "La empresa natural de Felipe II y la primera expedición científica en suelo americano: la creación del modelo expedicionario renacentista." In J. Martínez Millán, ed., *Felipe II (1527–1598): Europa y la monarquía católica,* 4: 39–59. Madrid: Parteluz, 1998.

———. "Francisco Hernández, Plinio del Nuevo Mundo: Tradición clásica, teoría nominal y sistema terminológico indígena en una obra renacentista." In Berta Ares Queija and Serge Gruzinski, eds., *Entre dos mundos: Fronteras culturales y agents mediadores,* 243–68. Seville: Escuela de Estudios Hispanoamericanos, 1997.

Caballero-Barnard, José Manuel, and Luis Coto y Maldonado. *El enamorado del volcán: Biografía artística del pintor Luis Coto.* Mexico City: Gobierno del Estado de México, Dirección de Turismo, 1977.

Cabello Valboa, Miguel. *Miscelánea antártica: Una historia del Peru antiguo.* Ca. 1586. Edited by the Instituto de Etnología. Lima: Universidad Nacional Mayor de San Marcos, Facultad de Letras, Instituto de Etnología, 1951.

Cabrera, Miguel. "Maravilla americana." 1756. Reproduced in Ernesto de la Torre Villar and Ramiro Navarro de Anda, eds., *Testimonios históricos guadalupanos* (Mexico City: Fondo de Cultura Económica, 1982), 494–528.

Cabrera y Quintero, Cayetano de. *Escudo de armas de México: Celestial protección de la Nueva España y de casi todo el Nuevo Mundo.* Mexico City: Joseph Bernardo de Hogal, 1746.

Cadalso, José de. *Cartas marruecas.* 1793. Edited by Manuel Camarero. Madrid: Castalia, 1985.

———. *Defensa de la nación española contra la Carta persiana LXXVII de Montesquieu (texto inédito).* Edited by Guy Mercadier. Toulouse: Institut d'études hispaniques, hispano-americaines et luso-brésiliennes, Université de Toulouse, 1979.

———. *Los eruditos a la violeta.* 1772. Edited by Nigel Glendinning. Salamanca: Anaya, 1967.

Cahill, David. "Colour by Numbers: Racial and Ethnic Categories in the Viceroyalty of Peru, 1532–1824." *Journal of Latin American Studies* 26 (1994): 325–46.

Calancha, Antonio de la. *Corónica [sic] moralizada del orden de San Augustín.* Barcelona: Pedro Lacavalleria, 1638.

Caldas, Francisco José de. "Ensayo sobre el estado de la geografía en el Virreino de Santa Fe de Bogotá con relación a la economía y al comercio (1808)." In J. Chenu, ed., *Francisco José de Caldas: Un peregrino de las ciencias.* Madrid: Historia 16, 1992.

———. *Obras completas.* Bogotá: Universidad Nacional de Colombia, 1966.

———. "Influjo del clima sobre los seres organizados (1808)." In *Obras completas de Francisco José de Caldas.* Bogotá: Universidad Nacional de Colombia, 1966.

Camacho, Arturo. *Album del tiempo perdido: Pintura Jalisciense del siglo XIX.* Guadalajara: El Colegio de Jalisco, 1997.

The Cambridge History of Science. Volume 4: *Eighteenth-Century Science.* Edited by Roy Porter, David C. Lindberg, and Ronald L. Numbers. New York: Cambridge University Press, 2003.

Camões, Luís de. *Os Lusíadas.* 1572. Lisbon: Publicações Europa-America, 1997.

Cannon, Susan F. *Science in Culture: The Early Victorian Period.* New York: Science History Publications, 1978.

Cañeque, Alejandro. *The King's Living Image: The Culture and Politics of Viceregal Power in Colonial Mexico.* New York: Routledge, 2004.

Cañizares-Esguerra, Jorge. *How to Write the History of the New World: Histories, Epistemologies, and Identities in the Eighteenth-Century Atlantic World.* Stanford, Calif.: Stanford University Press, 2001.

———. *Puritan Conquistadors: Iberianizing the Atlantic, 1550–1700.* Stanford, Calif.: Stanford University Press, 2006.

———. "Renaissance Mess(*tizaje*): What Mexican Indians Did to Titian and Ovid." *New Centennial Review* 2, 1 (2002): 267–76

———. "La utopía de Hipólito Unanue: Comercio, naturaleza, y religión en el Perú." In Marcos Cueto, ed., *Saberes andinos: Ciencia y tecnología en Bolivia, Ecuador y Perú,* 91–108. Lima: Instituto de Estudios Peruanos, 1995.

Capel, Horacio. *Geografía y matemáticas en la España del siglo XVIII.* Barcelona: Oikus-Tau, 1982.

Capel, Horacio, Joan-Eugeni Sánchez, and Omar Moncada, eds. *De Palas a Minerva: La formación científica y la estructura institucional de los ingenieros militares en el siglo XVIII.* Barcelona: Consejo Superior de Investigaciones Científicas y Ediciones Serbal, 1988.

Capmany, Antonio de [aka Pedro Fernández]. "Comentarios sobre el doctor festivo y maestro de los Eruditos a la Violeta, para desengaño de los españoles que leen poco y malo." 1773. In Julián Marías, *La España posible en tiempo de Carlos III,* 130–52. 2d ed. Madrid: Planeta, 1988.

Cárdenas, Juan de. *Primera parte de los problemas, y secretos marauillosos de las Indias.* Mexico: Pedro Ocharte, 1591.

———. *Problemas y secretos maravillosos de las Indias.* 1591. Madrid: Alianza, 1988.

Carpio, Manuel. "México." 1849. In *Antología poesía romántica mexicana,* ed. María del Carmen Millán, 40–48. Mexico City: Libro Mex, 1957.

———. *Poesías.* Edited by José Joaquín Pesado. Mexico City: Imprenta de M. Murgia dirigida por A. Contreras, 1849.

Carson, James Taylor. "Ethnogeography and the Native American Past." *Ethnohistory* 49 (2002): 269–88.

Castro, Concepción. *Campomanes: estado y reformismo ilustrado.* Madrid: Alianza, 1996.

Cervantes, Fernando. *The Devil in the New World: The Impact of Diabolism in New Spain.* New Haven, Conn.: Yale University Press, 1994.

Chaplin, Joyce E. *Subject Matter: Technology, the Body, and Science on the Anglo-American Frontier, 1500–1676.* Cambridge, Mass.: Harvard University Press.

Chappe d'Auteroche, Jean-Baptiste. *Voyage en Californie pour l'observation du passage de Vénus sur le disque du soleil, le 3 juin 1769; contenant les observations de ce phénomene, & la description historique de la route de l'auteur à travers le Mexique.* Paris: Charles-Antoine Jombert, 1772.

Chenu, Jeanne, ed. *Francisco José de Caldas: Un peregrino de las ciencias.* Madrid: Historia 16, 1992.

Chiaramonte, José Carlos, ed. *La Ilustración en el Río de la Plata: Cultura eclesiástica y cultura laica durante el Virreinato.* Buenos Aires: Punto Sur, 1989.

Cígala, Francisco Ignacio. *Cartas at Ilmo, y Rmo P. Mro. F. Benito Gerónymo Feyjoó Montenegro. Carta Segunda.* Mexico City: Biblioteca Mexicana, 1760.

Cisneros, Diego. *Sitio, naturaleza y propiedades de la ciudad de México: Aguas y vientos a que está sujeta; y tiempos del año; necesidad de su conocimiento para el exercicio de la medicina su incertidumbre y dificultad sin él de la astrología assi para la curación como para los prognósticos.* Mexico City: En casa del Bachiller Ioan Blanco de Alcaçar, 1618.

Clark, Stuart. *Thinking with Demons: The Idea of Witchcraft in Early Modern Europe.* Oxford: Oxford University Press, 1997.

Clément, Jean-Pierre. "L'apparition de la presse periodique en Amérique espagnole: Le cas du 'Mercurio Peruano.'" In *L'Amérique espagnole à l'epoque des lumières: Tradition, innovation, reprèsentation. Colloque franco-espagnol du CNRS, 18–20 Septembre 1986*, 173–86. Paris: CNRS, 1987.

———. "De los nombres de plantas." In Fermín del Pino, ed., *Ciencia y contexto histórico en las expediciones ilustradas a América*, 141–71. Madrid: Consejo de Investigaciones Científicas, 1988.

Clendinnen, Inga. *Ambivalent Conquests: Maya and Spaniard in Yucatan, 1517–1570.* Cambridge: Cambridge University Press, 1987.

Clusius, Carolus. *Admiranda narratio, fida tamen, de commodis et incolarvm ritibvs Virginiae.* Frankfurt a/M: T. de Bry, 1590.

———. *Aliqvot notae in Garciae aromatum historiam. . . . .* Antwerp: Christophori Plantini, 1582.

———. *Aromentum et simplicium aliquot medicamentorum apud Indos nascientum historia ante biennium quidem Lusitanica linqua per dialogos conscripta.* Abridged edition of Orta's *Coloquios.* Antwerp: Christophori Plantini, 1567.

———. *Rariorum aliquot stirpium per Hispanias obseruatarum historia: libris duobus expressa.* Antwerp: Christophori Plantini, 1576.

———. *Simpliciis medicamentis ex Occidentali India delatis.* Edition of Monardes's *Dos libros.* Antwerp: Christophori Plantini, 1574.

Cobo, Bernabé. *History of the Inca Empire.* Translated and edited by Roland Hamilton. Austin: University of Texas Press, 1979.

———. *Obras del Padre Cobo.* Edited by Francisco Mateos. 2 vols. Biblioteca de Autores Españoles, 91–92. Madrid: Atlas, 1956.

Coffin, David. *The Villa d'Este at Tivoli.* Princeton, N.J.: Princeton University Press, 1960.

———. *The Villa in the Life of Renaissance Rome.* Princeton, N.J.: Princeton University Press, 1979.

Columbus, Christopher. "Tercer Viage." In *The Four Voyages of Columbus: A History in Eight Documents, Including Five by Christopher Columbus in the Original*

Spanish with English Translations, ed. and trans. Cecil Jane, 2: 29–47. 2 vols. New York: Dover, 1988.

Comito, Terry. *The Idea of the Garden in the Renaissance*. New Brunswick, N.J.: Rutgers University Press, 1978.

Conquistador anónimo. *Relación de algunas cosas de la Nueva España y de la gran ciudad de Temistitán, México*. 1st ed. 1556. Translated by Francisco de la Maza. Edited by Federico Gómez de Orozco. Mexico City: Porrúa, 1961.

Contreras, Carlos, and Guillermo Mira. "Transferencia de tecnología minera de Europa a los Andes." In Antonio Lafuente et al., eds., *Mundialización de la ciencia y cultura nacional*, 235–49. Madrid: Doce Calles, 1993.

Cope, R. Douglas. *The Limits of Racial Domination: Plebeian Society in Colonial Mexico City, 1660–1720*. Madison: University of Wisconsin Press, 1994.

Cortés, Martín. *The arte of nauigation conteyning a compendious description of the sphere . . . translated out of the Spanyshe into Englyshe by Richard Eden*. London, 1561.

Corteseão, Armando. *Cartografia e cartógrafos portugueses dos séculos XV e XVI*. Vol. 2. Lisbon: Edição da "Seara nova," 1935.

Cotarelo y Mori, Emilio. *Iriarte y su época*. Madrid: Estudio Tipográfico "Sucesores de Rivadeneyra," 1897.

Craib, Raymond B. "A Nationalist Metaphysics: State Fixations, National Maps, and the Geo-Historical Imagination in Nineteenth-Century Mexico." *Hispanic American Historical Review* 82 (2002): 33–68.

Cruz, Ramón de la. *Sainetes*. Edited by Francisco Lafarga. Madrid: Cátedra, 1990.

Cueto, Marcos, ed. *Saberes andinos: Ciencia y tecnología en Bolivia, Ecuador y Peru*. Lima: Instituto de Estudios Peruanos, 1995.

Dandelet, Thomas James. *Spanish Rome, 1500–1700*. New Haven, Conn.: Yale University Press, 2001.

Darnall, Margaretta, and Mark S. Weil. "Il sacro bosco di Bomarzo: Its Sixteenth-Century Literary and Antiquarian Context." *Journal of Garden History* 4 (1984): 1–94.

Daston, Lorraine, and Katherine Park. *Wonders and the Order of Nature, 1150–1750*. New York: Zone Books, 1998.

Dávila Padilla, Agustín. *Historia de la fundación y discurso de la provincia de Santiago de México, de la Orden de Predicadores, por las vidas de sus varones insignes, y casos notables de Nueva España*. 1596. Brussels: Ivan de Meerveque, 1625.

De la Cruz, Martín, and Juan Badiano. *The Badianus Manuscript, Codex Barberini, Latin 241, Vatican Library: An Aztec Herbal of 1552*. Translated by Emily Walcott Emmart. Baltimore: Johns Hopkins University Press, 1940.

De la Puente, Juán. *De la conveniencia de las dos monarquías católicas la del iglesia romana y la del imperio español, y defensa de la procedencia de los Reyes Católicos de España a todos los reyes del mundo*. Madrid: Imprenta Real, 1612.

De Vos, Paula. "The Art of Pharmacy in Seventeenth- and Eighteenth-Century Mexico." PhD diss., University of California, Berkeley, 2001.

———. "An Herbal El Dorado: The Quest for Botanical Wealth in the Spanish Empire." *Endeavour* 27 (2003): 117–21.

Dear, Peter. *Revolutionizing the Sciences: European Knowledge and Its Ambitions, 1500–1800*. Princeton, N.J.: Princeton University Press, 2001.

Delumeau, Jean. *Paradise: The Garden of Eden in Myth and Tradition*. Translated by Matthew O'Connell. New York: Continuum, 1995.

Dettelbach, Michael. "Global Physics and Aesthetics Empire: Humboldt's Physical Portrait of the Tropics." In David Philip Miller and Peter Hanns Reill, eds., *Visions of Empire: Voyages, Botany, and Representation of Nature*, 258–92. Cambridge: Cambridge University Press, 1996.

———. "Humboldtian Science." In N. Jardine, J. A. Secord, and E. C. Spary, eds., *Cultures of Natural History*, 287–304. Cambridge: Cambridge University Press, 1996.

Díaz y de Ovando, Clementina. "La voz del paisaje en la literature mexicana, 1867–1912." In Xavier Moyssén Echeverría et al., *José María Velasco: Homenaje*, 265–303. Mexico City: Universidad Nacional Autónoma de México, 1989.

Díez Borque, José María. *La vida española en el Siglo de Oro según los extranjeros*. Barcelona: Serbal, 1990.

Dorantes de Carranza, Baltazar. *Sumaria relación de las cosas de Nueva España*. 1604. Edited by José María de Agreda y Sánchez. Mexico: Imprenta Museo Nacional, 1902.

Drayton, Richard. *Nature's Government: Science, Imperial Britain, and the "Improvement" of the World*. New Haven, Conn.: Yale University Press, 2000.

Ecott, Tim. *Vanilla: Travels in Search of the Ice Cream Orchid*. New York: Grove, 2004.

Edel, Abraham. *Aristotle and His Philosophy*. Chapel Hill: University of North Carolina Press, 1982.

Eguiara y Eguren, Juan José de. "Panegírico de la Virgen de Guadalupe." 1756. Reproduced in Ernesto de la Torre Villar and Ramiro Navarro de Anda, eds., *Testimonios históricos guadalupanos* (Mexico City: Fondo de Cultura Económica, 1982), 480–93.

El Censor: Obra Periódica comenzada a publicar en 1781 y terminada en 1787. Edited by José Miguel Caso González. Oviedo: Universidad de Oviedo, Instituto Feijoó de Estudios del siglo XVIII, 1989.

Elguero, Francisco. "La Virgen de Guadalupe." 1890s. In Octaviano Valdés, ed., *Poesia neoclásica y académica*. Mexico: Universidad Nacional Autónoma de México, 1946.

Elliott, John H. *The Old World and the New, 1492–1650*. Cambridge: Cambridge University Press, 1970.

———. *Spain and Its World, 1500–1700*. New Haven, Conn.: Yale University Press, 1989.

Encina, Juan de la. *El paisajista: José María Velasco (1840–1912)*. Mexico City: El Colegio de México, 1943.

Engstrand, Iris H. W. *Spanish Scientists in the New World: The Eighteenth-Century Expeditions*. Seattle: University of Washington Press, 1981.

Eriksson, Gunnar. *The Atlantic Vision: Olaus Rudbeck and Baroque Science*. Canton, Mass.: Science History Publications, 1994.

Evans, William McKee. "From the Land of Canaan to the Land of Guinea: The Strange Odyssey of the 'Sons of Ham.'" *American Historical Review* 85 (1980): 15–43.

Farías, Manuel Ignacio. *Eclypse del divino sol causada por la interposición de la inmaculada luna.* Mexico City: Maria de Rivera, 1742.

Farías Nuñez, Marcelo. *Tras El Dorado vegetal: José Celestino Mutis y la Real Expedición Botánica del Nuevo Reino de Granada.* Seville: Diputación de Sevilla, 1994.

Farriss, Nancy. *Maya Society Under Colonial Rule.* Princeton, N.J.: Princeton University Press, 1984.

Fernández de Echeverría y Veytia, Mariano. *Baluartes de Mexico.* Ca. 1778. Mexico City: Alejandro Valdés, 1820.

Fernández de Enciso, Martín. *Suma de geographia que trata de todas las partidas y provincias del mundo en especial de las Indias, [y] trata largamente del arte del marear juntamente con la espera en romance: con el regimie[n]to del sol y del norte, agora nueuamente emendada de algunos defectos q[ue] tenia en la impressio[n] passada.* Seville: Juan Cromberger, 1530.

Fernández de Oviedo, Gonzalo. *La historia general de las Indias.* Seville: Juan Cromberger, 1535.

———. *Sumario de la natural historia de las Indias.* Edited by Manuel Ballesteros. Madrid: Historia 16, 1986.

Feuillée, Louis. *Journal des observations physiques, mathématiques et botaniques sur les côtes orientales de l'Amérique Méridionale et dans les Indes Occidentales, depuis l'année 1707 jusque en 1712.* 2 vols. Paris: Pierre Giffart, 1714.

———. *Journal des observations physiques, mathématiques et botaniques sur les côtes orientales de la Amérique Méridionale et aux Indes Occidentales, et dans un autre voiage fait par le même ordre à la Nouvelle Espagne et aux isles de l'Amérique.* Paris: Jean Matiette, 1725.

Findlen, Paula, ed. *Athanasius Kircher: The Last Man Who Knew Everything.* New York: Routledge, 2004.

Flora de la Real Expedición Botánica del Nuevo Reino de Granada. 47 vols. Madrid: Ediciones de Cultura Hispánica, 1954–.

Florencia, Francisco. "Estrella del norte de Mexico." 1688. Reproduced in Ernesto de la Torre Villar and Ramiro Navarro de Anda, eds., *Testimonios históricos guadalupanos* (Mexico City: Fondo de Cultura Económica, 1982), 359–99.

Forner, Juan Pablo. *Discurso sobre el modo de escribir y mejorar la historia de España: Informe fiscal.* 1796. Edited by François Lopez. Barcelona: Labor, 1973.

———. *Oración apologética por la España y su mérito literario.* Madrid: Imprenta Real, 1786.

Freedberg, David. *The Eye of the Lynx: Galileo, His Friends, and the Beginnings of Modern Natural History.* Chicago: University of Chicago Press, 2002.

Freedman, Paul. *The Image of the Medieval Peasant as Alien and Exemplary.* Stanford, Calif.: Stanford University Press, 1999.

Fresquet Febrer, José Luis. *La experiencia americana y la terapéutica en los "Secretos de Chirugia" (1567) de Pedro Arias de Benavides.* Valencia: Instituto de Historia de la Ciencia y Documentación de la Universidad de Valencia, 1993.

Frézier, Amédée François. *Relation du voyage de la Mer du Sud aux côtes du Chily et Pérou fait pendant les années 1712, 1713 et 1714.* Paris: Chez Nyon. 1732.

Frías Núñez, Marcelo. "El té de Bogotá: Un intento de alternativa al té de China." In *Nouveau Monde el rennouveau de l'histoire naturelle,* ed. Marie-Cécile Bénassy-Berling, vol. 3, 201–19. Paris: Presses de la Sorbonne Nouvelle, 1994.

———. *Tras el Dorado vegetal: José Celestino Mutis y la Real Expedición Botánica del Nuevo Reino de Granada (1783–1808).* Seville: Diputación Provincial de Sevilla, 1994.

Gage, Thomas. *A New Survey of the West India's: or, The English American, his Travail by Sea and Land.* 2d enl. ed. London: Printed by E. Cotes and sold by John Sweeting, 1655.

Gándara, Miguel Antonio de. *Apuntes sobre el bien y el mal de España.* 1759. Edited by Jacinta Macías Delgado. Madrid: Instituto de Estudios Fiscales, 1988.

García Camarero, Ernesto, and Enrique García Camarero, eds. *La polémica de la ciencia española.* Madrid: Alianza, 1970.

García Carcel, Ricardo. *La leyenda negra: Historia y opinión.* Madrid: Alianza, 1981.

García de Céspedes, Andrés. *Regimiento de navegación.* Madrid, 1606.

García de la Vega, José Vicente. *Discurso crítico sobre el uso de las lagartijas, como específico contra muchas enfermedades.* Mexico City: Felipe de Zúñiga y Ontiveros, 1782.

García, Gregorio. *Origen de los indios del Nuevo Mundo.* 1729. Edited by Franklin Pease. Mexico City: Fondo de Cultura Económica, 1981.

García Tapia, Nicolás. *Técnica y poder en Castilla durante los siglos XVI y XVII.* Salamanca: Junta de Castilla y León, 1989.

García y Cubas, Antonio. *Memoria para servir a la carta general de la república mexicana.* Mexico City: Andrade y Escalante, 1861.

Gay, Peter. *Schnitzler's Century: The Making of Middle Class Culture, 1815–1914.* New York: Norton, 2002.

Geneva, Ann. *Astrology and the Seventeenth-Century Mind: William Lilly and the Language of the Stars.* Manchester: Manchester University Press, 1995.

Gerbi, Antonello. *The Dispute of the New World: The History of a Polemic, 1750–1900.* Translated by Jeremy Moyle. Rev. enl. ed. Pittsburgh: University of Pennsylvania Press, 1973. Originally published as *La disputa del Nuovo Mondo: Storia di una polemica, 1750–1900* (Milan: R. Ricciardi, 1955).

———. *La natura delle Indie nove: da Cristoforo Colombo a Gonzalo Fernández de Oviedo.* Milan: R. Ricciardi, 1975. Translated by Jeremy Moyle as *Nature in the New World: from Christopher Columbus to Gonzalo Fernández de Oviedo* (Pittsburgh: University of Pittsburgh Press, 1985).

Gibson, Charles. *The Aztecs Under Spanish Rule: A History of the Indians of the Valley of Mexico, 1519–1810.* Stanford, Calif.: Stanford University Press, 1964.

———, ed. *The Black Legend: Anti-Spanish Attitudes in the Old World and New.* New York: Knopf, 1971.

Giron, Nicole. "El paisajismo de Ignacio Manuel Altamirano." In Xavier Moyssén Echeverría et al., *José María Velasco: Homenaje,* 233–63. Mexico City: Universidad Nacional Autónoma de México, 1989.

Glacken, Clarence J. *Traces on the Rhodian Shore: Nature and Culture in Western Thought from Ancient Times to the End of the Eighteenth Century*. Berkeley: University of California Press, 1967.

Glick, Thomas F. "Science and Independence in Latin America (with Special Reference to New Granada)." *Hispanic American Historical Review* 71 (1991): 307–34.

González Bueno, Antonio, and Alberto Gomis Blanco. *Los naturalistas españoles en el Africa hispana (1860–1936)*. Madrid: Organismo Autónomo Parques Nacionales, 2002.

González Bueno, Antonio, ed. *Expedición botánica al virreinato del Peru (1777–1788)*. Barcelona: Lunwerg, 1988.

González Casanova, Pablo. *Misoneismo y modernidad cristiana en el siglo XVIII*. Mexico City: Colegio de México, 1958.

González Claverán, Virginia. *La expedición científica de Malaspina en Nueva España (1780–1794)*. Mexico City: Colegio de México, 1988.

González, Enrique. "El rechazo de la Universidad de México a las reformas ilustradas (1763–1777)." *Estudios de historia social y económica de América* (Alcalá) 7 (1991): 94–114.

Goodman, David C. *Power and Penury: Government, Technology and Science in Philip II's Spain*. Cambridge: Cambridge University Press, 1988.

——. "The Scientific Revolution in Spain and Portugal." In Roy Porter and Mikulás Teich, eds., *The Scientific Revolution in National Context*. Cambridge: Cambridge University Press, 1992.

Gould, Stephen Jay. *The Mismeasure of Man*. New York: Norton, 1981.

Grafton, Anthony. *Cardano's Cosmos: The Worlds and Work of a Renaissance Astrologer*. Cambridge, Mass: Harvard University Press, 1999.

——. *New Worlds, Ancient Texts: The Power of Tradition and the Shock of Discovery*. Cambridge, Mass: Harvard University Press, 1995.

Graham, Richard, ed. *The Idea of Race in Latin America*. Austin: University of Texas Press, 1990.

Greenfield, Amy Butler. *A Perfect Red: Empire, Espionage, and the Quest for the Color of Desire*. New York: HarperCollins, 2005.

Gruzinski, Serge. *L'Amérique de la conquête: Peinte par les Indiens du Mexique*. Paris: UNESCO/Flammarion, 1991.

——. *La colonisation de l'imaginaire: Sociétés indigènes et occidentalisation dans le Mexique espagnol XVIe–XVIIIe siècle*. Paris: Gallimard, 1988.

Guamán Poma de Ayala, Felipe. *Nueva crónica y buen gobierno*. Edited by John Murra, Rolena Adorno, and Jorge Urioste. Madrid: Historia 16, 1987.

Hakluyt, Richard. *The principal navigations, voyages, traffiques and discoveries of the English nation: made by sea or over-land to the remote and farthest distant quarters of the earth at any time within the compass of these 1600 yeares*. 1589. 2d ed. 1598–1600. Reprinted in 12 vols. Glasgow: James MacLehose & Sons, 1903–5.

Hale, Charles A. *The Transformation of Liberalism in Late Nineteenth-Century Mexico*. Princeton, N.J.: Princeton University Press, 1990.

Hall, A. Rupert. *The Scientific Revolution, 1500–1800: The Formation of the Modern Scientific Attitude*. London: Longmans, 1954.

Hall, Stuart. "When Was 'The Post-Colonial'? Thinking at the Limit." In Ian

Chambers and Lidia Curti, eds., *The Post-Colonial Question: Common Skies, Divided Horizons*, 242–60. London: Routledge, 1996.

Hamblyn, Richard. *The Invention of Clouds: How an Amateur Meteorologist Forged the Language of the Skies*. New York: Farrar, Straus & Giroux, 2001.

Hamy, E. T. *Joseph Dombey, médicin, naturaliste, archéologue, explorateur du Pérou, du Chili et du Brésil (1778–85)*. Paris: E. Guilmoto, 1905.

Hanke, Lewis. *Aristotle and the American Indians: A Study in Race Prejudice in the Modern World*. London: Hollis & Carter, 1959.

Hannaford, Ivan. *Race: The History of an Idea in the West*. Baltimore: Johns Hopkins University Press, 1996.

Harris, Steven J. "Jesuit Scientific Activity in Overseas Missions, 1540–1773." *Isis* 96 (2005): 71–79.

Heredia, José María. "El tecocalli de Cholula." In Emmanuel Carballo, ed., *Poesía mexicana del siglo XIX*. Mexico City: Díogenes, 1984.

Hernández de Alba, Gonzalo. *Quinas Amargas: El sabio Mutis y la discusión naturalista del siglo XVIII*. Bogotá: Academia de Historia de Bogotá and Tercer Mundo, 1991.

Hernández, Francisco *Antigüedades de Nueva España*. Ca. 1580. Madrid: Historia 16, 1986.

———. *Rerum medicarum Novae Hispaniae thesaurus. . . .* Rome: Vitalis Mascardi. 1651.

Herr, Richard. *The Eighteenth-Century Revolution in Spain*. Princeton, N.J.: Princeton University Press, 1958.

Herrera y Tordesillas, Antonio de. *Historia general de los hechos de los castellanos en las islas y tierra firme del Mar Océano, 1492–1531*. 9 vols. in 4. Madrid: Imprenta Real, 1601–15. See also the edition in 7 vols. by Antonio Ballesteros y Beretta (Madrid: Academia de la Historia, 1934–57).

Hillgarth, J. N. *The Mirror of Spain, 1500–1700: The Formation of a Myth*. Ann Arbor: University of Michigan Press, 2000.

Hirshman, Albert O. *The Passions and the Interests: Political Arguments for Capitalism Before Its Triumph*. Princeton, N.J.: Princeton University Press, 1977.

Hobbes, Thomas. *Leviathan. Or the Matter, Forme and Power of a Commonwealth Ecclesiasticall and Civil*. 1651. Edited by Michael Oakeshott. New York: Collier Books, 1962.

Hoberman, Louisa Schell. "Enrico Martínez: Printer and Engineer." In D. G. Sweet and G. B. Nash, eds., *Struggle and Survival in Colonial America*, 331–46. Berkeley: University of California Press, 1981.

Hont, Istvan, and Michael Ignatieff, eds. *Wealth and Virtue: The Shaping of Political Economy in the Scottish Enlightenment*. Cambridge: Cambridge University Press, 1983.

Hooykaas R. *Humanism and the Voyages of Discovery in Sixteenth-Century Portuguese Science and Letters*. Amsterdam: North-Holland, 1979.

Horne, Thomas. *The Social Thought of Bernard Mandeville: Virtue and Commerce in Early Eighteenth-Century England*. New York: Columbia University Press, 1978.

Huddleston, Lee Eldrige. *Origins of the American Indians: European Concepts, 1492–1729*. Austin: University of Texas Press, 1967.

Select Bibliography

Hudson, Nicholas. "From 'Nation' to 'Race': The Origin of Racial Classification in Eighteenth-Century Thought." *Eighteenth-Century Studies* 29 (1996): 247–64.

Hulme, Peter. "Tales of Distinction: European Ethnography and the Caribbean." In Stuart B. Schwartz, ed., *Implicit Understandings: Observing, Reporting, and Reflecting on the Encounters Between Europeans and Other Peoples in the Early Modern Era*, 157–97. Cambridge: Cambridge University Press, 1994.

Humboldt, Alexander von. *Voyage de Humboldt et Bonpland: Voyage aux régions équinoxiales du nouveau continent*. 1805–30. 30 vols. Reprint. Amsterdam: Theatrum Orbis Terrarum; New York: Da Capo Press, 1971–73.

Ishikawa, Chiyo, ed. *Spain in the Age of Exploration, 1492–1819*. Seattle: Seattle Art Museum; Lincoln: University of Nebraska Press, 2004.

Israel, Jonathan Irvine. *Race, Class, and Politics in Colonial Mexico, 1610–1670*. London: Oxford University Press, 1975.

Jacobsen, Nils. *Mirages of Transition: The Peruvian Altiplano, 1780–1930*. Berkeley: University of California Press, 1993.

Jalisco Genio y Maestría. Monterrey, Mexico: Museo de Arte Contemporareno de Monterrey, 1994.

Joaquin Clausell y los ecos del impresionismo en México. Mexico City: Museo Nacional de Arte, 1995.

Johns, Michael. *The City of Mexico in the Age of Díaz*. Austin: University of Texas Press, 1997.

Jovellanos, Melchor Gaspar de. *Obras sociales y políticas*. Edited by Patricio Peñalver Simó. Madrid: Publicaciones Españolas, 1962.

Juárez Medina, Antonio. *Las reediciones de obras de erudición de los siglos XVI y XVII durante el siglo XVIII español*. Frankfurt a/M: Lang, 1988.

Juderías, Julián. *La leyenda negra: Estudios acerca del concepto de España en el extranjero*. 9th ed. Barcelona: Araluce, 1943.

Jurkevich, Gayana. *In Pursuit of the Natural Sign: Azorín and the Poetics of Ekphrasis*. London: Associated University Presses, 1999.

Kagan, Richard L. "*Arcana Imperii*: Maps, Knowledge and Power at the Court of Philip IV." In Felipe Pereda and Fernando Marias, eds., *El atlas del rey planeta: La "descripción de España y de las costas y puertos de sus reinos" de Pedro Texeira (1634)*. Madrid: Nerea, 2002.

———. "Clio and the Crown: Writing Histories in Habsburg Spain." In Richard L. Kagan and Geoffrey Parker, eds., *Spain, Europe, and the Atlantic World: Essays in Honour of John H. Elliott*, 73–100. Cambridge: Cambridge University Press, 1995.

———. *Urban Images of the Hispanic World, 1493–1793*. New Haven, Conn.: Yale University Press, 2000.

Kamen, Henry. *Empire: How Spain Became a World Power, 1492–1763*. New York: HarperCollins, 2003.

Kelley, Donald R. *The Human Measure. Social Thought in the Western Legal Tradition*. Cambridge, Mass.: Harvard University Press, 1990.

Kellner, L. *Alexander von Humboldt*. New York: Oxford University Press, 1963.

Kelly, Franklin. *Frederic Edwin Church and the National Landscape*. Washington, D.C.: Smithsonian Institution Press, 1988.

Kemys, Lawrence. *A Relation of the Second Voyage to Guiana*. London: Thomas Dawson, 1596.

Kircher, Athanasius. *Magneticum naturae regnum*. Rome: Ignacio de Lazaris, 1667.

Knobel, E. B. "On Friederick de Houtman's Catalogue of Southern Stars and the Origin of the Southern Constellations." *Monthly Notices of the Royal Astronomical Society* 76 (1917): 414–32.

Koerner, Joseph Leo. *Caspar David Friedrich and the Subject of Landscape*. New Haven, Conn.: Yale University Press, 1990.

Koerner, Lisbet. *Linnaeus: Nature and Nation*. Cambridge, Mass.: Harvard University Press, 1999.

Kupperman, Karen Ordahl. "Climate and Mastery of the Wilderness in Seventeenth-Century New England." In David D. Hall and David Grayson Allen, eds., *Seventeenth-Century New England*, 3–37. Boston: Colonial Society of Massachusetts, 1984.

——. "Fear of Hot Climates in the Anglo-American Colonial Experience." *William and Mary Quarterly*, 3d ser., 41 (1984): 215–40.

——. "The Puzzle of the American Climate in the Early Colonial Period." *American Historical Review* 87 (1982): 1262–89.

——, ed. *America in European Consciousness, 1493–1750*. Chapel Hill, N.C.: Published for the Institute of Early American History and Culture, Williamsburg, Va., by the University of North Carolina Press, 1995.

L'Escola d'Olot: J. Berga, J. Vayreda, M. Vayreda. Barcelona: Fundació La Caixa, 1993.

La Condamine, Charles-Marie de. *Journal du voyage fait par ordre du roi à l'Equateur*. Paris: Imprimerie Royale, 1751.

——. *Lettre à Madame*** sur l'emeute populaire excitée en la ville de Cuenca au Pérou le 29 d'Aôut. 1730*. Paris, 1746.

La Porte, Joseph de. *Le voyageur françois*. 42 vols. Paris: Cellot, 1768–95.

Lafaye, Jacques. *Quetzalcoatl et Guadalupe: La formation de la consciente nacionales au Mexique (1531–1813)*. Paris: Gallimard, 1974.

Lafuente, Antonio, and Antonio Mazuecos. *Los caballeros del punto fijo: Ciencia, política y aventura en la expedición geodésica hispanofrancesa al virreinato del Perú en el siglo XVIII*. Barcelona: Serbal, 1987.

Lafuente, Antonio, and José Luis Peset. "Las academias militares y la inversión en ciencia en la España ilustrada (1750–1760)." *Dynamis* 2 (1982): 193–209.

Lafuente, Antonio, and José Sala Catalá, eds. *Ciencia colonial en América*. Madrid: Alianza, 1992.

Lafuente, Antonio, Alberto Elena, and M. L. Ortega, eds. *Mundialización de la ciencia y cultura nacional*. Madrid: Doce Calles, 1993.

Lagos, Licio. *El paisaje mexicano en la Colección Lagos*. Mexico City: Fondo de Cultura Económica, 1971.

Lamb, Ursula. *Cosmographers and Pilots of the Spanish Maritime Empire*. Aldershot, UK: Variorum, 1995.

Lampillas, Francisco Xavier. *Ensayo histórico-apologético de la literatura española contra las opiniones preocupadas de algunos escritores modernos italianos*. Vol. 7. Zaragoza: Blas Miedes, 1782–89.

Langley, Lester D. *The Americas in the Age of Revolution, 1750–1850*. New Haven, Conn.: Yale University Press, 1996.

Lanning, John Tate. *Academic Culture in the Spanish Colonies*. London: Oxford University Press, 1940.

——. *The Eighteenth-Century Enlightenment in the University of San Carlos de Guatemala*. Ithaca, N.Y.: Cornell University Press, 1956.

——. *The Royal Protomedicato: The Regulation of the Medical Professions in the Spanish Empire*. Edited by John Jay TePaske. Durham, N.C.: Duke University Press, 1985.

Laqueur, Thomas. *Making Sex: Body and Gender from the Greeks to Freud*. Cambridge, Mass: Harvard University Press, 1990.

Las Casas, Bartolomé de. *Apologética historia sumaria*. Edited by Edmundo O'Gorman. 2 vols. Mexico City: Universidad Nacional Autónoma de México, 1967.

Lavallé, Bernard. *Las promesas ambiguas: Ensayos sobre el criollismo colonial en los Andes*. Lima: Pontificia Universidad Católica del Perú, Instituto Riva Agüero, 1993.

Lazzaro, Claudia. *The Italian Renaissance Garden: From the Conventions of Planting, Design and Ornament to the Grand Gardens of Sixteenth-Century Central Italy*. New Haven, Conn.: Yale University Press, 1990.

León Pinelo, Antonio de. *Epítome de la biblioteca oriental y occidental, naútica y geográfica*. Madrid: Juan González, 1629.

——. *El paraíso en el Nuevo Mundo: Comentario apologético, historia natural y peregrina de las Indias Occidentales*. Edited by Raúl Porras Barrenechea. 2 vols. Lima: Torres Aguirre, 1943.

Leonard, Irving A. *Baroque Times in Old Mexico*. Ann Arbor: University of Michigan Press, 1959.

——. *Don Carlos de Sigüenza y Góngora*. Berkeley: University of California Press, 1929.

Lertora Mendoza, Celina A. "Introducción de las teorías newtonianas en el Río de la Plata." In Antonio Lafuente et al., eds., *Mundialización de la ciencia y cultura nacional*, 307–23. Madrid: Doce Calles, 1993.

Lindberg, David C., and Robert S. Westman, *Reappraisals of the Scientific Revolution*. Cambridge: Cambridge University Press, 1991.

Linnaeus, Carl. "On the Increase of the Habitable Earth." In *Select Dissertations from the Amoenitates Academicae* (1781), 1: 71–127. Translated by F. J. Brand. 2 vols. New York: Arno Press, 1977.

Litvak, Lily. *El tiempo de los trenes: El paisaje español en el arte y la literatura del realismo, 1849–1918*. Barcelona: Serbal, 1991.

Lizana, Bernardo de. *Devocionario de Nuestra Señora de Izamal, y conquista espiritual de Yucatán*. 1633. I have used the edition by Felix Jiménez Villalba, *Historia de Yucatán* (Madrid: Historia 16, 1988).

Llano Zapata, José Eusebio de. *Epítome cronológico o idea general del Perú. Crónica inédita de 1776*. Edited by Victor Peralta. Madrid: Editorial Mapfre, 2005.

——. *Memorias, histórico, físicas, critico, apologéticas de la América Meridional [tres tomos]*. Edited by Ricardo Ramírez, Antonio Garrido, Luis Millones Figueroa, Victor Peralta, and Charles Walker. Lima: Instituto Francés de Estudios Andinos; Pontificia Universidad Católica del Perú; Universidad Mayor de San Marcos, 2005.

———. *Memorias histórico-físicas-apologéticas de la América Meridional.* 1770. Lima: Imprenta y Librería de San Pedro, 1904.

Lockhart, James. *The Nahuas After the Conquest: A Social and Cultural History of the Indians of Central Mexico, Sixteenth Through Eighteenth Centuries.* Stanford, Calif.: Stanford University Press, 1992.

Lopez, François. "Comment l'Espagne eclairée inventa le Siècle d'Or." In *Hommage des hispanistes Français à Noël Salomon*, 517–25. Barcelona: Laia, 1979.

———. *Juan Pablo Forner et la crise de la conscience espagnole au XVIIIIe siècle.* Bordeaux: Bibliothèque de l'École des hautes études hispaniques, 1976.

López de Gómara, Francisco. *Historia de las Indias y conquista de México.* Zaragoza: Agustín Millan, 1552.

López-Ocón, Leoncio, and Carmen María Pérez-Montes Salmerón, eds. *Marcos Jiménez de la Espada (1831–1898): Tras la senda de un explorador.* Madrid: CSIC, 2000.

López-Ocón, Leoncio, and Miguel Angel Puig-Samper. "Los condicionamientos políticos de la Comisión Científica del Pacífico: Nacionalismo e hispanoamericanismo en la España bajoisabelina (1854–1868)." *Revista de Indias* 47 (1987): 667–82.

López Piñero, José María. *Ciencia y técnica en la sociedad española de los siglos XVI y XVII.* Barcelona: Labor Universitaria, 1979.

López Piñero, José María, and Francisco Calero. *"De Pulvere febrífugo Occidentalis Indiae" (1663) de Gaspar Caldera de Heredia y la introducción de la quina en Europa.* Valencia: Instituto de Historia de la Ciencia y Documentación de la Universidad de Valencia, 1992.

López Piñero, José María, and José Pardo Tomás. *La influencia de Francisco Hernández (1515–1587) en la constitución de la botánica y la materia médica moderna.* Valencia: Instituto de Historia de la Ciencia y Documentación de la Universidad de Valencia, 1994.

———. *Nuevos materials y noticias sobre la "Historia de las plantas de Nueva España" de Francisco Hernández.* Valencia: Instituto de Historia de la Ciencia y Documentación de la Universidad de Valencia, 1994.

López Piñero, José María, and María Luz López Terrada. *La influencia española en la introducción en Europa de las plantas americanas, 1493–1623.* Valencia: Instituto de Historia de la Ciencia y Documentación de la Universidad de Valencia, 1997.

Lozoya, Xavier. *Plantas y luces en México: La real expedición científica a Nueva España (1787–1803).* Barcelona: Serbal, 1984.

Lupher, David A. *Romans in the New World: Classical Models in Sixteenth-Century Spanish America.* Ann Arbor: University of Michigan Press, 2003.

MacCormack, Sabine. *Religion in the Andes: Vision and Imagination in Early Colonial Peru.* Princeton, N.J.: Princeton University Press, 1991.

MacDougall, E. B., ed. *Medieval Gardens.* Washington, D.C.: Dumbarton Oaks, 1986.

Maclean, Ian. *The Renaissance Notion of Woman: A Study in the Fortunes of Scho-*

lasticism and Medical Science in European Intellectual Life. Cambridge: Cambridge University Press, 1980.

Malagón, Javier, and José M. Ots Capdequí. *Solórzano y la política indiana.* 1965. 2d ed. Mexico City: Fondo de Cultura Económica, 1983.

Malpica Diosdado, Joseph Francisco de. *Alexipharmaco de la salud, antídoto de la enfermedad, favorable dietético instrumento de la vida. Dissertacion medico-moral, que trata del ayuno, y accidentes, que escusan de él, y que hacen licito el uso de las carnes á los enfermos, y valetudinarios.* Mexico City: Colegio Real de San Ildefonso, 1751.

Maltby, William S. *The Black Legend in England: The Development of Anti-Spanish Sentiment, 1558–1660.* Durham, N.C.: Duke University Press, 1971.

Manthorne, Katherine Emma. *Tropical Rennaisance: North American Artists Exploring Latin America, 1839–1879.* Washington, D.C.: Smithsonian Institution Press, 1989.

Maravall, José Antonio. *Antiguos y modernos: La idea de progreso en el desarrollo inicial de una sociedad.* Madrid: Sociedad de Estudios y Publicaciones, 1966.

———. *Antiguos y modernos: Visión de la historia de la idea de progreso hasta el Renacimiento.* 2d ed. Madrid: Alianza, 1986.

———. *La cultura del barroco.* 6th ed. Madrid: Ariel, 1996.

———. "El sentimiento de nación en el siglo XVIII: La obra de Forner" (1967). In Maria Carmen Iglesias, ed., *Estudios de la historia del pensamiento español, siglo XVIII,* 42–60. Madrid: Mondadori España, 1991.

Martín, Luis. *The Intellectual Conquest of Peru: The Jesuit College of San Pablo, 1568–1767.* New York: Fordham University Press, 1968.

Martínez, Enrico. *Repertorio de los tiempos y historia natural desta Nueva España.* Mexico City: Enrico Martínez, 1606.

Martínez Ruiz, Enrique, ed. *Felipe II, la ciencia y la técnica.* Madrid: Actas, 1999.

Masdeu, Juan Francisco. *Historia crítica de España y de la cultura española.* 20 vols. Madrid: Sancha, 1783–1805.

Matienzo, Joseph Giral. "Aprovación." In José Vicente García de la Vega, *Discurso crítico sobre el uso de las lagartijas, coma específico contra muchas enfermedades.* Mexico City: Felipe de Zúñiga y Ontiveros, 1782.

Maza, Francisco de la. *Enrico Martínez, cosmógrafo e impresor de Nueva España.* Mexico City: Sociedad Mexicana de Geografía y Estadística, 1943.

———. *El guadalupanismo mexicano.* 1953. Mexico City: Fondo de Cultura Económica y Secretaría de Educación Pública, 1984.

McClellan, James E., III. *Colonialism and Science: Saint Domingue in the Old Regime.* Baltimore: Johns Hopkins University Press, 1992.

McCook, Stuart. *States of Nature: Science, Agriculture, and Environment in the Spanish Caribbean, 1760–1940.* Austin: University of Texas Press, 2002.

Mendoza Vargas, Héctor, ed. *México a través de los mapas.* Mexico City: Plaza y Valdés, 2000.

Menes-Llaguno, Juan Manuel. *Bartolomé de Medina: Un sevillano pachuquero.* Pachuca, Mexico: Universidad Autónoma del Estado de Hidalgo, 1989.

Mestre, Antonio. *Mayans y la España de la Ilustración.* Madrid: Instituto de España and Espasa-Calpe, 1990.

Mexico una visón de su paisaje: A Landscape Revisited. Washington, D.C: Smithsonian Institution Traveling Exhibition Service and Mexican Cultural Institute, 1994.

México y sus alrededores: Colección de monumentos, trajes y paisajes. Dibujados al natural y litografiados por los artistas mexicanos C. Castro, J. Campillo, L. Auda y G. Rodríguez. Mexico: Establecimiento litográfico de Decaen, 1855–56.

Mier Noriega y Guerra, José Servando Teresa de. "Sermón predicado en la Colegiata el 12 de diciembre de 1794." In id., *Obras completas: El heterodoxo guadalupano.* Edited by Edmundo O'Gorman. 2 vols. Mexico City: Universidad Nacional Autónoma de México, 1981.

Millán, María del Carmen. *El paisaje en la poesía mexicana.* Mexico City: Imprenta Universitaria, 1952.

Miller, Angela. *The Empire of the Eye: Landscape Representation and American Cultural Politics.* Ithaca, N.Y.: Cornell University Press, 1996.

Miller, David Philip, and Peter Hanns Reill, eds. *Visions of Empire: Voyages, Botany, and Representation of Nature.* Cambridge: Cambridge University Press, 1996.

Miller, Robert R. *For Science and National Glory: The Spanish Scientific Expedition to America, 1862–1866.* Norman: University of Oklahoma Press, 1968.

Minguet, Charles. *Alexandre de Humboldt: Historien et géographe de l'Amérique espagnole, 1799–1804.* Paris: Maspero, 1969.

Mínguez, Victor. *Los reyes distantes: Imágenes del poder en México virreinal.* Castelló de la Plana, Spain: Universidad Jaume I, 1995.

———. *Los reyes solares: Iconografía astral de la monarquía hispánica.* Castelló de la Plana, Spain: Universidad Jaume I, 2001.

Monardes, Nicolás. *Dos libros: El Vno Qve Trata De Todas Las Cosas que traen de nuestras Indias Occidentales, que siruen al vso de la Medicina, y el otro que trata de la Piedra Bezaar, y de la Yerua Eseuerçonera.* Seville: Heruando Díaz and Alonso Ejcriuano, 1565–69.

Montesquieu, Charles de Secondat, baron de. *Lettres persannes,* no. 78 (1721). In Charles Gibson, ed., *The Black Legend: Anti-Spanish Attitudes in the Old World and the New.* New York: Knopf, 1971.

Morel-Fatio, Alfred. "Comment la France a connu et compris l'Espagne depuis le moyen âge jusqu'à nos jours." In id., *Études sur l'Espagne,* 1st ser., 1–114. Paris: F. Vieweg, 1888.

Moreno, Roberto, ed. *Linneo en Mexico.* Mexico City: Universidad Nacional Autónoma de Mexico, 1989.

Morner, Magnus. *Race Mixture in the History of Latin America.* Boston: Little, Brown, 1967.

Moyssén Echeverría, Xavier. *José María Velasco.* Mexico City: Fondo de la Plástica Mexicana, 1991.

Moyssén Echeverría, Xavier, et al. *José María Velasco: Homenaje.* Mexico City: Universidad Nacional Autónoma de México, 1989.

Mukerji, Chandra. *Territorial Ambitions and the Gardens of Versailles.* Cambridge: Cambridge University Press, 1997.

Muldoon, James. *The Americas in the Spanish World Order: The Justification for Conquest in the Seventeenth Century.* Philadelphia: University of Pennsylvania Press, 1994.

Select Bibliography

Mundy, Barbara E. *The Mapping of New Spain: Indigenous Cartography and the Maps of the Relaciones Geográficas*. Chicago: University of Chicago Press, 1996.

Muñoz, Juan Bautista. *Juicio del tratado de la educación del M. R. P. D. Cesareo Pozzi*. Madrid: Joachim Ibarra, 1778.

Muro, Luis. "Bartolomé de Medina, introducción del beneficio de patio en Nueva España." *Historia Mexicana* 13 (1964): 517–31.

Murra, John V. *Formaciones económicas y políticas del mundo andino*. Lima: Instituto de Estudios Peruanos, 1975.

Mutis, José Celestino. *Escritos científicos de Don José Celestino Mutis*. Edited by Guillermo Hernández de Alba. 2 vols. Bogotá: Instituto Colombiano de Cultura Hispánica, 1983.

Nader, Helen. *The Mendoza Family in the Spanish Renaissance, 1350–1550*. New Brunswick: Rutgers University Press, 1979.

Navarro Brotóns, Victor. "Las ciencias en la España del siglo XVII: El cultivo de las disciplinas físico-matemáticas." *Arbor* 153 (1996): 197–252.

———. "The Reception of Copernicus in Sixteenth-Century Spain: The Case of Diego de Zúñiga." *Isis* 86 (1995): 52–78.

Navarro Brótons, Victor, and Enrique Rodríguez Galdeano. *Matemáticas, cosmología y humanismo en la España del siglo XVI: Los Comentarios al segundo libro de la Historia natural de Plinio de Jerónimo Muñoz*. Valencia: Instituto de estudios documentales e históricos sobre la ciencia, 1998.

Newman, William. *Gehennical Fire: The Lives of George Starkey, an American Alchemist in the Scientific Revolution*. Cambridge, Mass.: Harvard University Press, 1994.

Newman, William, and Lawrence M. Principe. *Alchemy Tried in the Fire: Starkey, Boyle, and the Fate of Helmontian Chymistry*. Chicago: University of Chicago Press, 2002.

Nicholas Hudson, "From 'Nation' to 'Race': The Origin of Racial Classification in Eighteenth-Century Thought." *Eighteenth-Century Studies* 29 (1996): 247–64.

Nicolson, Malcom. "Alexander von Humboldt, Humboldtian Science and the Origins of the Study of Vegetation." *History of Science* 25 (1987): 167–94.

Nieremberg, Juan Eusebio. *Curiosa filosofía y tesoro de maravillas de la naturaleza, examinadas en varias questiones naturales. Contienen historias muy notables. Averígüense secretos y Problemas de la naturaleza, con Filosofía nueva. Explícanse lugares dificultosos de Escritura. Obra muy útil, no solo para los curiosos, sino para doctos Escriturarios, Filósofos, y médicos*. Barcelona: Pedro Lacavelleria, 1644.

———. *Historia natvrae, maxime peregrinae, libris XVI distincta. In quibus rarissima naturae arcana . . . etiam cum proprietatibus medicinalibus, describuntur*. Antwerp: Moreti, 1635.

———. *Ocvlta filosofía de la simpatía, y antipatía de las cosas. Artificio de la naturaleza, y noticia natural del mundo. Y segunda parte de la Curiosa filosofía: contiene historias notables, averiguasen muchos secretos, y problemas de la naturaleza*. Barcelona: Pedro Lacavalleria, 1645.

Nieto Olarte, Mauricio. *Remedios para el imperio: Historia natural y la apropriación del Nuevo Mundo*. Bogotá: Instituto Colombiano de Antropología e Historia, 2000.

Nirenberg, David. *Communities of Violence: Persecution of Minorities in the Middle Ages.* Princeton, N.J.: Princeton University Press, 1996.

Noriega, Tomas. "El Ahuehuete" (1877). In *La Naturaleza: Periódico científico de la Sociedad Mexicana de Historia Natural* 4 (1877–79): 35–40.

Novak, Barbara. *Nature and Culture: American Landscape and Painting, 1825–1875.* New York: Oxford University Press, 1980.

Núñez, Frías. "El té de Bogotá: Un intento de alternativa al té de China." In Marie-Cécile Bénassy, Jean-Pierre Clément, Francisco Pelayo, Miguel Angel Puig-Samper, eds., *Nouveau Monde et renouveau de l'histoire naturelle,* 3: 201–19. 3 vols. Paris: Presses de la Sorbonne nouvelle, 1986–94.

Olson, Richard. *The Emergence of the Social Sciences, 1642–1792.* New York: Twayne, 1993.

Orozco y Berra, Manuel. *Apuntes para la historia de la geografía en México.* Mexico City: Francisco Díaz de León, 1881.

Orta, García de. *Coloquios dos simples e drogas e coisas medicinais da India e de algumas frutas achadas nella onde se tratam algu[m]as cousas tocantes a mediçina, practica, e outras cousas boas pera saber.* Goa: Ioannes de Endem, 1563.

Ortega Cantero, Nicolás. "La institución Libre de Enseñanza y el entendimiento del paisaje madrileño." *Anales de Geografía de la Universidad Complutense* 6 (1986): 81–98.

Ortega Reyes, Manuel. "El gigante de la flora mexicana o sea el sabino de Santa María de Tule del Estado de Oaxaca" (1882). *La Naturaleza: Periódico científico de la Sociedad Mexicana de Historia Natural* 6 (1882–84): 110–14.

Osasunasco, Desiderio de. *Observaciones sobre la preparación y usos del chocolate.* Mexico City: Felipe de Zúñiga y Ontiveros, 1789.

Osorio Romero, Ignacio. *La luz imaginaria: Epistolario de Atanasio Kircher con los novohispanos.* Mexico City: Universidad Nacional Autónoma de México, 1993.

Osorio y Peralta, Didaco. *Principia medicinae epitome, et totius humanis corporis fabrica seu ex microcosmi armonia divinium, germen.* Mexico City: Heredes viduae Bernardo Calderon, 1685.

Othón, Manuel José. "Las montañas épicas" (1902). In id., *Paisaje,* ed. Manuel Calvillo, 77–80. Mexico City: Universidad Nacional Autónoma de México, 1944.

———. *Paisaje.* Edited by Manuel Castillo. Mexico City: Universidad Nacional Autónoma de México, 1944.

Outram, Dorinda. *The Enlightenment.* Cambridge: Cambridge University Press, 1995.

Pagden, Anthony. *The Fall of Natural Man: The American Indian and the Origins of Comparative Ethnology.* Rev. ed. Cambridge: Cambridge University Press, 1986.

———. *Spanish Imperialism and the Political Imagination.* New Haven, Conn.: Yale University Press, 1990.

Paisaje y otros pasajes mexicanos del siglo XIX en la Colección del Museo Soumaya. Mexico City: Museo Soumaya, 1998.

Pardo Tomás, José, and María Luz López Terrada. *Las primeras noticias sobre plantas americanas en las relaciones de viajes y crónicas de Indias (1493–1553).*

Valencia: Instituto de historia de la ciencia y documentación de la Universidad de Valencia, 1993.

Park, Katherine. "The Organic Soul." In C. Schmitt, Q. Skinner, E. Kessler, and J. Kraye, eds., *The Cambridge History of Renaissance Philosophy*, 464–84. Cambridge: Cambridge University Press, 1988.

Pascual, Ricardo. *El botánico José Quer (1695–1764): Primer apologista de la ciencia española*. Valencia: Cátedra e Instituto de Historia de la Medicina, 1970.

Payno, Manuel. *Panorama de México*, Vol. 5 of *Obras completas*. Edited by Boris Rosen Jélomer. Mexico City: Consejo Nacional para la Cultura y las Artes, 1999.

Paz, Octavio. *Sor Juana or the Traps of Faith*. Cambridge, Mass.: Harvard University Press, 1988.

Pegaza, Joaquín Arcadio. *Murmurios de la selva*. Mexico City: Francisco Díaz León, 1887.

Pena, María del Carmen. *Pintura de paisaje e ideología: La generación del 98*. Madrid: Taurus, 1982.

Peña, Rafael Angel de la. "Prólogo." In Joaquín Arcadio Pegaza, *Murmurios de la selva*. Mexico City: Francisco Díaz León, 1887.

Peña Montenegro, Alonso de la. *Itinerario para párrocos de indios*. Madrid: Joseph Fernández de Buendía, 1668.

Perdices Blas, Luis. *La economía política de la decadencia de Castilla en el siglo XVII: Investigaciones de los arbitristas sobre la naturaleza y causa de la riqueza de las naciones* Madrid: Síntesis, 1996.

Pérez de Salazar, Francisco. *Biografía de D. Calos de Sigüenza y Góngora*. Mexico City: Antigua Imprenta de Murguía, 1928.

Pesado, José Joaquín. "A la santísima Virgen de Guadalupe" (1840s). In Octaviano Valdés, ed., *Poesia neoclásica y académica*. Mexico City: Universidad Nacional Autónoma de México, 1946.

Peset, Mariano, and José Luis Peset. "La renovación universitaria." In Manuel Selles et al., eds., *Carlos III y la ciencia de la Ilustración*, 143–55.

Phelan, John Leddy. *The Millennial Kingdom of the Franciscans in the New World: A Study of the Writings of Gerónimo de Mendieta, 1525–1604*. Berkeley: University of California Press, 1956.

———. "Neo-Aztecism in the Eighteenth Century and the Genesis of Mexican Nationalism." In Stanley Diamond, ed., *Culture in History: Essays in Honor of Paul Radin*. New York: Columbia University Press, 1960.

Pimentel, Juan. *La física de la monarquía: Ciencia y política en el pensamiento colonial de Alejandro Malaspina (1754–1810)*. Madrid: Doce Calles, 1998.

———. "The Iberian Vision: Science and Empire in the Framework of a Universal Monarchy, 1500–1800." In Roy MacLeod, ed., *Nature and Empire: Science and the Colonial Enterprise. Osiris*, 2d ser., 15 (2000): 17–30.

———. *Testigos del mundo: Ciencia, literatura y viajes en la Ilustración*. Madrid: Marcial Pons Historia, 2003.

Pino, Fermín del, and Angel Guirao. "Las expediciones ilustradas y el estado español." *Revista de Indias* 47 (1987): 380–83.

Pocock, J. G. A. *The Machiavellian Moment: Florentine Political Thought and the Atlantic Republican Tradition*. Princeton, N.J.: Princeton University Press, 1975.

———. *Virtue, Commerce, and History*. Cambridge: Cambridge University Press, 1985.

Pomar, Jaume Honorat. *El códice de Jaume Honorat Pomar (c. 1550–1606): Plantas y animales del viejo mundo y de América*. Edited by José María López Piñero. Valencia: Ajuntament de València, 2000.

Popkin, Richard Henry. *Isaac La Peyrère (1596–1676): His Life, Work, and Influence*. Leiden: Brill, 1987.

Pratt, Mary Louise. *Imperial Eyes: Travel Wrting and Transculturation*. New York: Routledge, 1992.

Prest, John M. *The Garden of Eden: The Botanic Garden and the Recreation of Paradise*. New Haven, Conn.: Yale University Press, 1981.

Prieto, Guillermo. "Chapoltepec." *Museo mexicano o Miscelanea pintoresca de amenidades curiosas e instructivas* 3 (1844): 212–16.

Pruna, Pedro M. "National Science in a Colonial Context: The Royal Academy of Sciences in Havana, 1861–1898." *Isis* 85 (1994): 412–26.

Puerto Sarmiento, Francisco Javier. *Ciencia de cámara: Casimiro Gómez Ortega (1741–1818): El científico cortesano*. Madrid: Consejo Superior de Investigaciones Científicas, 1992.

———. *La ilusión quebrada: Botánica, sanidad y política científica en la España ilustrada*. Barcelona: Serbal, 1988.

Puig-Samper, Miguel Angel. *Crónica de una expedición romántica al Nuevo Mundo: La Comisión Científica del Pacífico (1862–1866)*. Madrid: CSIC, 1988.

Puig-Samper, Miguel Angel, and Francisco Pelayo. "Las expediciones botánicas al nuevo mundo durante el siglo XVIII: Una aproximación historiográfica." In Daniela Soto Arango et al., eds., *La Ilustración en América colonial*, 55–65. Madrid: Doce Calles, CSIC, and Colciencias, 1995.

Purchas, Samuel. *Pvrchas his pilgrimage: or, Relations of the world and the religions obserued in al ages and places discouered, from the creation vnto this present*. London: William Stansby for Henry Fetherstone, 1617.

Quint, David. *Epic and Empire: Politics and Generic Form from Virgil to Milton*. Princeton, N.J.: Princeton University Press, 1993.

Quintana, José Miguel. *La astrología en la Nueva España en el siglo XVII (de Enrico Martínez a Sigüenza y Góngora)*. Mexico City: Talleres Gráficos, 1969.

Ramírez, Fausto. "Introduction: On José María Velasco's Artistic Achievements." In María Elena Altamirano Piolle, *National Homage: José María Velasco (1840–1912)*, 1: 25–34. 2 vols. Mexico City: Amigos del Museo Nacional de Arte, 1993.

Raynal, Guillaume-Thomas-François, abbé. *Histoire philosophique et politique des établissements et du commerce des Européens dans les deux Indes*. 1770. 3d ed. 10 vols. Geneva: Jean-Léonard Pellet, 1781. Translated by J. O. Justamond as *Philosophical and Political History of the Settlements and Trade of the Europeans in the East and West Indies* (3d ed., 8 vols., London: W. Strahan & T. Cadell, 1783).

Reyes, Alfonso. *El paisaje en la poesía mexicana del siglo XIX*. Mexico City: Viuda de F. Díaz de León, 1911.

Rico, Francisco. *El pequeño mundo del hombre: Varia fortuna de una idea en las letras españolas.* Madrid: Castalia, 1970.

Riera, Juan. "Médicos y cirujanos extranjeros de cámara en España del siglo XVIII." *Cuadernos de historia de la medicina española* 14 (1975): 87–104.

Riva Palacio, Vicente, et al. *México a través de los siglos: Historia general y completa del desenvolvimiento social, político, religioso, militar, artístico, científico y literario de México desde la antigüedad más remota hasta la época actual.* 5 vols. Mexico City: Ballescá, 1887–89.

Robertson, Donald. *Mexican Manuscript Painting of the Early Colonial Period: The Metropolitan Schools.* New Haven, Conn.: Yale University Press, 1959.

Roca, Luis de la. "El bosque de Chapultepec." In *México y sus alrededores: Colección de monumentos, trajes y paisajes. Dibujados al natural y litografiados por los artistas mexicanos C. Castro, J. Campillo, L. Auda y G. Rodríguez,* 25. Mexico: Establecimiento litográfico de Decaen, 1855–56.

Rocha, Diego Andrés. *El origen de los indios.* 1681. Edited by José Alcina Franch. Madrid: Historia 16, 1988.

Rodríguez, Diego. *Discurso etheorológico del nuevo cometa.* Mexico City: Viuda de Bernardo Calderón, 1652.

Rodríguez O., Jaime E. "The Emancipation of America." *American Historical Review* 105 (2000): 131–52.

———. *La independencia de la América española.* Mexico City: Fondo de Cultura Económica and Colegio de Mexico, 1996.

Rodríguez Prampolini, Ida, ed. *La crítica de arte en México en el siglo XIX.* 3 vols. Mexico City: Universidad Nacional Autónoma de México, 1997.

Romano, David. *La ciencia hispanojudía.* Madrid: Mapfre, 1992.

Romero de Terreros, Manuel. "Los descubridores del paisaje mexicano." *Artes de México* 5 (1959).

———. *Paisajistas mexicanos del siglo XIX.* Mexico City: Imprenta Universitaria, 1943.

Romero, Félix. "El Popocateptl." *Oaxaca Herald,* April 27, 1907. Reproduced in Fildalfo Figueroa, *10 poetas oaxaqueños del siglo XIX.* Oaxaca: "El Flechador del Sol" de Apoco, 1988.

Rosenthal, Earl. "The Invention of the Columnar Device of Emperor Charles V at the Court of Burgundy in Flanders in 1516." *Journal of the Warburg and Courtauld Institutes* 36 (1973): 198–230.

———. "Plus Ultra, Non Plus Ultra, and the Columnar Device of Emperor Charles V." *Journal of the Warburg and Courtauld Institutes* 34 (1971): 204–28.

Rothschild, Emma. *Economic Sentiments: Adam Smith, Condorcet and the Enlightenment.* Cambridge, Mass.: Harvard University Press, 2001.

Rubial García, Antonio. *Una monarquía criolla: La provincia augustiniana de Mexico en el siglo XVII.* Mexico City: Consejo Nacional para la Cultura y las Artes, 1990.

Sahagún, Bernardino de. *Historia General de las cosas de Nueva España.* Ca. 1570. Edited by Angel María Garibay. 4 vols. Mexico City: Porrúa, 1956.

Sala Catalá, José. *Ciencia y técnica en la metropolización de América.* Madrid: Doce Calles, 1994.

Salazar y Solana, Javier Pérez de. *José María Velasco y sus contemporáneos: Una muestra de la pintura academica nacional mexicana a fines del siglo XIX y principios del siglo XX, época de J. M. Velasco.* Monterrey: Perpal, 1982.

Saldaña, Juan José "Ilustración, ciencia y técnica en América." In Daniela Soto Arango et al., eds., *La Ilustración en América colonial*, 19–49. Madrid: Doce Calles, CSIC, and Colciencias, 1995.

Salinas y Córdova, Buenaventura de. *Memorial de las historias del Nuevo Mundo Piru: Méritos y excelencias de la ciudad de Lima, cabeza de sus ricos y estendidos reynos y el estado presente en el que se hallan.* Lima: Gerónimo de Contreras, 1630. I have also used the 1957 edition published by the Universidad Mayor de San Marcos, Lima.

———. *Memorial, informe, y manifiesto del P.F. Bvenaventvra de Salinas y Cordova, de la Orden de S. Francisco, Letor Iubilado, Calificador del Consejo de la Santa General Inquisicion, Padre de la Prouincia de los Doz Apostoles de Lima, y Comissario General de las de la Nueua-España: al Rey Nvestro Señor, en su Real, y Supremo Consejo de las Indias.* Rome, 1646.

San Vicente, Juan Manuel de. *Exacta descripción de la magnífica corte mexicana, cabeza del nuevo americano mundo.* Cadiz, 1768. Reproduced in *Anales del Museo Nacional de Antropología de Mexico*, 3d ser., 5 (1913).

Sánchez, Miguel. *Imagen de la Virgen María madre de Dios de Guadalupe: milagrosamente aparecida en la ciudad de México: celebrada en su historia, con la profecía del capítulo doze del Apocalipsis.* 1648. Reproduced in Ernesto de la Torre Villar and Ramiro Navarro de Anda, eds., *Testimonios históricos guadalupanos* (Mexico City: Fondo de Cultura Económica, 1982), 152–281.

Sánchez-Blanco Parody, Francisco. *El absolutismo y las luces en el reinado de Carlos III.* Madrid: Marcial Pons, 2002.

———. *Europa y el pensamiento español del siglo XVIII.* Madrid: Alianza, 1991.

Sandman, Alison. "Mirroring the World: Sea Charts, Navigation, and Territorial Claims in Sixteenth-Century Spain." In Pamela H. Smith and Paula Findlen, eds., *Merchants and Marvels: Commerce, Science, and Art in Early Modern Europe*, 83–108. New York: Routledge, 2002.

Sarrailh, Jean. *La España Ilustrada de la segunda mitad del siglo XVIII.* Mexico City: Fondo de Cultura Económica, 1957.

———. "Voyageurs français au XVIIIe siècle: De l'abbé de Vayrac à l'abbé Delaporte." *Bulletin Hispanique* 36, no. 1 (1934).

Sartorius, Carl Christian. *Mexico: Landscapes and Popular Sketches.* London: Trübner, 1859.

Scaliger, Giulio Cesare. *Exotericarum Exercitationum liber quintus decimus de subtilitate ad Hieronymum Cardanum.* Paris: Michaelis Vascosani, 1557.

Schiebinger, Londa. *The Mind Has No Sex? Women in the Origins of Modern Science.* Cambridge, Mass.: Harvard University Press, 1989.

Schiebinger, Londa, and Claudia Swan, eds. *Colonial Botany: Science, Commerce, and Politics in the Early Modern World.* Philadelphia: University of Pennsylvania Press, 2005.

Schmidt, Benjamin. *Innocence Abroad: The Dutch Imagination and the New World, 1570–1670.* Cambridge: Cambridge University Press, 2001.

Schmidt-Nowara, Christopher. *Empire and Antislavery: Spain, Cuba, and Puerto Rico, 1833–1874.* Pittsburgh: University of Pittsburgh Press, 1999.

Schroeder, Susan. "Looking Back at the Conquest: Nahua Perceptions of Early Encounters from the Annals of Chimalpahin." In Eloise Quiñonez Keber et al., eds., *Chipping Away on Earth: Studies in Prehispanic and Colonial Mexico in Honor of Arthur J. O. Anderson and Charles E. Dibble.* Lancaster, Calif.: Labyrinthos, 1994.

Seed, Patricia. "'Are These Not Also Men?': The Indians' Humanity and Capacity for Spanish Civilization." *Journal of Latin American Studies* 25 (1993): 629–52.

———. *Ceremonies of Possession in Europe's Conquest of the New World, 1492–1640.* Cambridge: Cambridge University Press, 1995.

———. "Social Dimensions of Race: Mexico City, 1753." *Hispanic American Historical Review* 62 (1982): 569–606.

Sellés, Manuel, José Luis Peset, and Antonio Lafuente, eds. *Carlos III y la ciencia de la Ilustración.* Madrid: Alianza, 1988.

Sempere y Guarinos, Juan. *Consideraciones sobre las causas de la grandeza y de la decadencia de la monarquía española.* Edited by Juan Rico Jiménez. Alicante: Instituto de Cultura "Juan Gil-Albert," 1998.

———. *Ensayo de una biblioteca española de los mejores escritores del reynado de Carlos III.* 6 vols. Madrid: Imprenta Real, 1785–89.

———. *Historia del luxo y de las leyes suntuarias de España.* 2 vols. Madrid: Imprenta Real, 1788.

Shapin, Steven. *The Scientific Revolution.* Chicago: University of Chicago Press, 1996.

Sicroff, Albert. *Los estatutos de limpieza de sangre: Controversias entre los siglos XV y XVII.* Madrid: Taurus, 1985.

Silva, Renán. *Prensa y revolución a finales del siglo XVIII.* Bogotá: Banco de la Republica, 1988.

Singleton, Charles S. "Stars over Eden." *Annual Report of the Dante Society* 75 (1975): 1–18.

Siraisi, Nancy G. *The Clock and the Mirror: Girolamo Cardano and Renaissance Medicine.* Princeton, N.J.: Princeton University Press, 1997.

———. *Medieval and Early Renaissance Medicine.* Chicago: University of Chicago Press, 1990.

Smith, Pamela H., and Paula Findlen, eds. *Merchants and Marvels: Commerce, Science, and Art in Early Modern Europe.* New York: Routledge, 2002.

Smoller, Laura. *History, Prophecy, and the Stars: The Christian Astrology of Pierre d'Ailly, 1350–1420.* Princeton, N.J.: Princeton University Press, 1994.

Solórzano Pereira, Juan. *Política indiana.* 1647. 5 vols. Biblioteca Autores Españoles, vol. 252–56 Madrid: Atlas, 1972.

Soto Arango, Diana. "La enseñanza ilustrada en las universidades de América colonial: Estudio historiográfico." In id. et al., eds., *La Ilustración en América colonial,* 91–119. Madrid: Doce Calles, CSIC, and Colciencias, 1995.

Soto Arango, Diana, Miguel Angel Puig-Samper, and Luis Carlos Arboleda, eds. *La Ilustración en América colonial.* Madrid: Doce Calles, CSIC, and Colciencias, 1995.

Soto, J. "Panorama de México: El Puente Nacional." *Museo mexicano o Miscelanea pintoresca de amenidades curiosas e instructivas* 2 (1843): 256–59.

Spary, E. C. "Political, Natural, and Bodily Economies." In N. Jardine, J. A. Secord, and E. C. Spary, eds., *Cultures of Natural History*, 178–96. Cambridge: Cambridge University Press, 1996.

———. *Utopia's Garden: French Natural History from Old Regime to Revolution*. Chicago: University of Chicago Press, 2000.

Steele, Arthur R. *Flowers for the King: The Expeditions of Ruiz and Pavón and the Flora of Peru*. Durham, N.C.: Duke University Press, 1964.

Steneck, Nicholas H. "Albert on the Psychology of Sense Perception." In James A. Weisheipl, ed., *Albertus Magnus and the Sciences*, 263–90. Toronto: Pontifical Institute of Mediaeval Studies, 1980.

———. *Science and Creation in the Middle Ages*. Notre Dame, Ind.: Notre Dame University Press, 1976.

Stepan, Nancy Leys. *"The Hour of Eugenics": Race, Gender and Nation in Latin America*. Ithaca, N.Y.: Cornell University Press, 1991.

———. *The Idea of Race in Science: Great Britain, 1800–1960*. Hamden, Conn.: Archon Books, 1982.

———. "Race and Gender: The Role of Analogy in Science." *Isis* 77 (1986): 261–77.

Stern, Steve J. *Peru's Indian Peoples and the Challenge of the Spanish Conquest: Huamanga to 1640*. Madison: University of Wisconsin Press, 1982.

Stocking, George W. *Race, Culture, and Evolution*. 1968. Chicago: University of Chicago Press, 1982.

Stoler, Ann Laura. *Race and the Education of Desire. Foucault's History of Sexuality and the Colonial Order of Things*. Durham, N.C.: Duke University Press, 1995.

Straet, Jan van der. *New Discoveries: The Sciences, Inventions, and Discoveries of the Middle Ages and the Renaissance as Represented in 24 Engravings Issued in the Early 1580's by Stradanus*. Norwalk, Conn.: Burndy Library, 1953.

———. *Nova reperta d'après Jean Stradanus; gravures de Philippe Galle*. Translated into French by André Stegmann, Guy Demerson, and Michel Reulos. Tours: O. Bienvault, 1977.

Strong, Roy. *The Renaissance Garden in England*. London: Thames & Hudson, 1979.

Suárez de Peralta, Juan. *Tratado del descubrimiento de las Indias (Noticias históricas de Nueva España)*. 1589. Mexico City: Secretaria de Educación Publica, 1949.

Tanner, Marie. *The Last Descendant of Aeneas: The Hapsburgs and the Mythic Image of the Emperor*. New Haven, Conn.: Yale University Press, 1993.

Ten, Antonio E. "Ciencia y universidad en la América hispana: La Universidad de Lima." In Antonio Lafuente and José Sala Catalá, eds., *Ciencia colonial en América*, 162–91. Madrid: Alianza, 1992.

Tenorio-Trillo, Mauricio. *Mexico at the World's Fairs: Crafting a Modern Nation*. Berkeley: University of California Press, 1996.

Terrall, Mary. *The Man Who Flattened the Earth: Maupertuis and the Sciences of the Enlightenment*. Chicago: University of Chicago Press, 2002.

Tietz, Manfred. "L'Espagne et *l'Histoire des deux Indes* de l'abbé Raynal." In Hans-Jürgen Lüsebrink and Manfred Tietz, eds., *Lectures de Raynal: L'Histoire des deux Indes en Europe et en Amérique au XVIIIe siècle. Actes du Colloque de Wolfenbüttel.* Oxford: Voltaire Foundation at the Taylor Institution, 1991.

Tooley, Marian J. "Bodin and the Medieval Theory of Climate." *Speculum* 28 (1953): 64–83.

Torquemada, Juan de. *Ia.[–IIIa.] parte de los veynte y vn libros rituales y monarchia yndiana: con el origen y guerras de los yndios occidentales de sus poblaçones descubrimiento conquista conuersion y otras cosas marauillosas de la mesma tierra: distribuydos en tres tomos.* 3 vols. Seville: Matthias Clavijo, 1615.

Torre Villar, Ernesto de la, and Ramiro Navarro de Anda, eds. *Testimonios históricos guadalupanos.* Mexico City: Fondo de Cultura Económica, 1982.

Torres Villaroel, Diego de. *Visiones y visitas de Torres con Don Francisco de Quevedo por la corte.* Edited by Sebold Russell. Madrid: Espasa-Calpe, 1966.

Trabulse, Elías. *La ciencia perdida: Fray Diego Rodríguez, un sabio del siglo XVII.* Mexico City: Fondo de Cultura Económica, 1985.

———. *El círculo roto.* Mexico City: Fondo de Cultura Económica, Secretaria de Educación Pública, 1984.

———. *Historia de la ciencia en México.* Abbreviated edition. Mexico City: Fondo de Cultura Económica y Consejo Nacional de Ciencia y Tecnología, 1994.

———. *José María Velasco: Un paisaje de la ciencia en México.* Toluca: Instituto Mexiquense de Cultura, 1992.

———. "La obra científica de Don Carlos de Sigüenza y Góngora, 1667–1700." In Antonio Lafuente and José Sala Catalá, eds., *Ciencia colonial en América,* 221–52. Madrid: Alianza, 1992.

Tres grandes maestros del paisaje decimonónico español: Jenaro Pérez Villaamil, Carlos de Haes, Aureliano de Beruete. Madrid: Centro Cultural del Conde Duque, 1990.

Unanue, Hipólito. "Disertación sobre el aspecto, cultivo, comercio y virtudes de la famosa planta del Perú nombrada Coca." *Mercurio Peruano,* nos. 372–77 (1794): 205–59.

———. "Geografía física del Perú." *Mercurio Peruano,* no. 4 (1792).

———. "Introducción a la descripción científica de las plantas del Perú." *Mercurio Peruano,* no. 2 (1791): 71.

———. *Observaciones sobre el clima de Lima y sus influencias en los seres organizados, en especial el hombre.* In Jorge Arias-Schreiber Pezet, ed., *Los ideólogos: Hipolito Unanue.* Vol. 8. Colección documental de la independencia del Peru, vols. 7–8. Lima: Comisión Nacional del Sesquicentenario de la Independencia del Perú, 1974.

Valadés, Diego. *Rhetorica Christiana ad concionandi et orandi usum ac commodata, utrissq facultatis exemplis suo loco insertis, quae quidem, es Indoroum maxme deprompta sunt histories, unde praeter doctrinam sumam quoque delectatio comparabitur.* Facsimile of the 1579 edition with parallel Spanish translation. Mexico City: Fondo de Cultura Económica, 1989.

Varela, Javier. *Jovellanos.* Madrid: Alianza, 1988.

Varey, Simon, ed. *The Mexican Treasury: The Writings of Dr. Francisco Hernández.* Translated by Rafael Chabrán, Cynthia L. Chamberlin, and Simon Varey. Stanford, Calif.: Stanford University Press, 2001.

Varey, Simon, Rafael Chabrán, and Dora B. Weiner, eds. *Searching for the Secrets of Nature: The Life and Works of Dr. Francisco Hernández*. Stanford, Calif.: Stanford University Press, 2000.

Vargas, Pedro Fermín de. "Memoria sobre la población del reino de Nueva Granada." Ca. 1790. In *Pensamientos políticos y memoria sobre la población del Nuevo Reino de Granada*. Bogotá: Biblioteca Popular de Cultura Colombiana, 1944.

Vaughan, Alden T. "From White Man to Redskin: Changing Anglo-American Perceptions of the American Indian." *American Historical Review* 87 (1982): 917–53.

Velasco, José María. "Anotaciones y observaciones al trabajo del señor Augusto Weismann sobre la transformación del ajolote mexicano en amblistoma." *La Naturaleza: Periódico científico de la Sociedad Mexicana de Historia Natural*, 1st ser., 5 (1880–81): 58–84. Reproduced in Trabulse, *José María Velasco* (Toluca: Instituto Mexiquense de Cultura, 1992), 271–91.

———. "Descripción, metamófosis y costumbres de una especie nueva del género Siredón encontrada en el Lago de Santa Isabel, cerca del Villa de Guadalupe Hidalgo, Valle de México" (1878). *La Naturaleza: Periódico científico de la Sociedad Mexicana de Historia Natural*, 1st ser., 4 (1877–79): 209–22. Reproduced in Trabulse, *José Maria Velasco* (Toluca: Instituto Mexiquense de Cultura, 1992), 229–45.

———. "Estudio anatómico de la circulación y respiración. Continuación de la memoria anterior" (1879). *La Naturaleza: Periódico científico de la Sociedad Mexicana de Historia Natural*, 1st ser., 4 (1877–79): 223–33. Reproduced in Trabulse, *José Maria Velasco* (Toluca: Instituto Mexiquense de Cultura, 1992), 229–45

———. "Estudios sobre la familia de las cacteas de México" (1870). In *La Naturaleza: Periódico científico de la Sociedad Mexicana de Historia Natural* 1 (1869–70): 201–3.

Velasco, José María, and Idelfonso Velasco. "Estudio sobre una nueva especie de falsa jalapa de Querétaro" (1870). *La Naturaleza: Periódico científico de la Sociedad Mexicana de Historia Natural* 1 (1869–70): 338–42. Reproduced in Trabulse, *José María Velasco* (Toluca: Instituto Mexiquense de Cultura, 1992), 226–28.

Venegas, Juan Manuel. *Compendio de la medicina o medicina práctica*. Mexico City: Felipe de Zúñiga y Ontiveros, 1788.

Vespucci, Amerigo. *Letters from a New World: Americo Vespucci's Discovery of America*. Translated by David Jacobson. Edited by Luciano Formisano. New York: Marsilio, 1992.

Vetancurt, Agustín de. *Teatro mexicano: Descripción breve de los svcessos exemplares, históricos, políticos, militares, y religiosos del Nuevo Mundo occidental de las Indias*. 2 vols. Mexico City: M. de Benavides, Viuda de Juan Ribera, 1697–98.

Vila, Pablo. "Caldas y los orígenes eurocriollos de la geobotánica." *Revista de la Academia Colombiana de Ciencias* 11 (1960): 16–20.

Vilar, Jean. *Literatura y economía: La figura satírica del arbitrista en el Siglo de Oro*. Madrid: Revista de Occidente, 1973.

Warner, Deborah J. *The Sky Explored: Celestial Cartography, 1500–1800*. New York: A. R. Liss, 1979.

Waterhouse, Edward. *A declaration of the state of the colony and affaires in Virginia. With a relation of the barbarous massacre in the time of peace and league, treach-*

Waterhouse, Edward. *A declaration of the state of the colony and affaires in Virginia. With a relation of the barbarous massacre in the time of peace and league, treacherously executed by the natiue infidels vpon the English, the 22 of March last.* London: G. Eld for Robert Mylbourne, 1622.

Weismann, August. "Transformación del ajolote mexicano en amblistoma" (1877). *La Naturaleza: Periódico científico de la Sociedad Mexicana de Histoira Natural,* 1st ser., 5 (1880–81): 31–57. Reproduced in Trabulse, *José María Velasco* (Toluca: Instituto Mexiquense de Cultura, 1992), 247–69.

Westfall, Richard S. *The Construction of Modern Science: Mechanisms and Mechanics.* New York: Wiley, 1971.

Whitaker, Arthur Preston. "The Elhuyar Mining Missions and the Enlightenment." *Hispanic American Historical Review* 31 (1951): 557–85.

Widdifield, Stacie G. *The Embodiment of the National in Late Nineteenth-Century Mexican Painting.* Tucson: University of Arizona Press, 1996.

Wobeser, Gisela von. "La pintura paisajista como testimonio de las haciendas en el siglo XIX." In Xavier Moyssén Echeverría et al., *José María Velasco: Homenaje,* 181–202. Mexico City: Universidad Nacional Autónoma de México, 1989.

Zavala, Jesús. *Manuel José Othón: El hombre y el poeta.* Mexico City: Impr. Universitaria, 1952.

Zea, Francisco Antonio. "Avisos de Hebephilo a los jóvenes de los dos colegios sobre la inutilidad de sus estudios presentes, necesidad de reformarlos, elección y buen gusto en los que deben abrazar." *Papel periódico de la ciudad de Santafé de Bogotá,* no. 9 (April 8, 1791).

Index

Index

Browne, Janet, 112
Bry, Theodore de, 28
Burke, Edmund, 5, 111

Caballe, Francisco Antonio, 126
Cabriada, Juan de, 33
Cadalso, José, 100–103
Calancha, Antonio de la, 50, 79, 85
Caldas, Francisco José de, 56, 113–16, 124–25
Calvo, Pedro, 159
Camões, Luis de, 18
Cañuelo, Luis, 107
Capellán, Juan, 33
Caramuel Lobkowitz, Juan, 33, 49
Cardano, Girolamo, 68, 73
Cárdenas, Juan de, 26, 69, 87–89
Carpio, Manuel, 161–62
cartography, history of in Mexico, 129–30, 190n1
Casa de Contratación, as clearinghouse of knowledge, 9, 19
Castro, Casimiro, 136, 159
Cavanilles, Antonio José, 96, 107
Censor (periodical), 109–10
Cerviño, Pedro Antonio, 126
Cessi, Federico, 29
Chabrán, Rafael, 41
Chaplin, Joyce E., 178n7
Chapman, Conrad Wise, 159
Chapultepec, as historical forest, 132–33, 134–36
Charles III, 57, 106, 132
Charles IV, 106
Charles V, 16; origins of his coat of arms, 170n1
Chaves, Alonso de, 18
Chaves, Jerónimo de, 19
Chimalpahin Quauhtlehuanitzin, Domingo de San Antón Muñón, 92
Church, Frederick Edwin, 150, 152–53
Cígala, Francisco Ignacio, 177n50
Cisneros, Diego, 90–91
Clausell, Joaquín, 164
Clavé, Pelegrín, 136, 137(fig)
Clusius, Carolus, 27–28, 171n28
Cobo, Bernabé, 85
coca leaves, 125
Codex Pomar, 35–36, 40
Cole, Thomas, 160
Colegio de Minería, 59

Colegio de Santa Cruz de Tlatelolco, 73, 174n2
Colegio Imperial, 39
Columbus, Christopher, 22, 117
comets, origins of, 49, 54
Cope, R. Douglass, 92
Copernicus, Nicolaus, 37–38
Corachán, Juan Bautista, 33
Córdoba, Bernardino de, 172n36
Corsali, Andrea, 70
Cortés, Hernán, 127, 133, 136, 140(fig), 142
Corzo, Carlos, 34
Coto, Luis, 145, 147, 150
Cristo del Sepulcro (Lord of Ameca), 139
Crusades, and epistemology, 3, 20
Cruz, Martín de la. *See* Badiano, Juan
Cruz, Sor Juana Inés de la, 140
Cuauhtemoc, 133
curse on descendants of Ham, 64, 65, 66, 85, 183n80
Cuvier, Georges, 151

Darwinian debates in Mexico, 150–52
Daston, Lorraine, 26
Dear, Peter, 45
Decaen, José Antonio (*México y sus alrededores*) 134–35, 155(fig)
Dee, John, 10
demonology: and idolatry, 94–95; and natural history, 26, 50–52
Denina, Carlo, 107
Descartes, René, 108
De Vos, Paula, 9
Díaz, Porfirio, 157, 161
Dioscorides, 8, 30, 36, 40, 41
Dorantes de Carranza, Baltasar, 26
Drake, Francis, 28
Drayton, Richard (*Nature's Government*), 27–28
Dumaine, Gregorio, 150
Duméril, Auguste, 151
Durkheim, Émile, 139

Eden, Richard, 72, 75
Egerton, Daniel Thomas, 159
Elliott, John H., 66
Ercilla, Alonso de, 18

Faber, Johannes, 29
Farías, Manuel Ignacio, 55
Farriss, Nancy, 46

Index

Petiver, James, 40
Philip II, 24, 28, 44; and alchemy, 35; and natural history, 35–36, 41; and gardens, 42; and *Relaciones geográficas*, 43; as cause of decline of Spain, 101
Philip III, 22, 127
Philip IV, 29, 64, 127
Pierson, Peter O'Malley, 41
Pigafetta, Antonio, 72
Pimentel, Juan, 22, 170n6
plague (*matlazahuatl*), causes of, 55
Plancius, Petrus (Pieter Platevoet), 80
Pliny, 18, 36
political economy: and commercial humanism, 98; as deductive science, 104, 110; as historical science, 99, 104, 108–9; and republican humanism, 98, 102
Pomar, Jaume Honorat, 35–36
Popocateptl. *See* volcanoes
Porte, Joseph la, 24
Portela Marco, Eugenio, 35
Poussin, Nicolas, 132
Pozzi, Cesareo Giuseppe, 99
Pozzo, Cassiano dal, 30
Pratt, Mary Louise, 153
Pre-Adamites, 66, 179n23
Prest, John M., 116
Prieto, Guillermo, 134
printing press. *See* scribal culture in Spain
Puente, Juan de la, 75
Puerto Sarmiento, Francisco, 58
Purchas, Samuel, 75

Quevedo, Francisco, 102
quinine, 122–23
Quint, David, 170n4

Raleigh, Sir Walter, 10, 127
Ramírez, Fausto, 156
Raynal, abbé, 24, 108, 109
Real Academia de Historia, 58
Recchi, Nardo Antonio, 28, 30
Relaciones geográficas, 9–10, 43
Risse, Guenter B., 41
Riva Palacio, Vicente (*México a través de los siglos*), 140–49, 156
Roca, Luís de la, 135
Rocha, Diego Andrés, 77
Rodríguez, Diego, 52–53, 81–82
Rodríguez Galván, Ignacio, 133

Rodríguez O., Jaime, 188n15
Rugendas, Johann Moritz, 154, 159

Sahagún, Bernardino, 73
Sala Catalá, José (*Ciencia y técnica*), 43–44
Saldaña, Juan José, 59
Salinas de Córdoba, Buenaventura, 64–65, 95
Salvador, Jaime, 33
Sánchez, Miguel, 52
Sandman, Alison, 173n45
San Vicente, Juan Manuel de, 126
Sartorius, Carl Christian (*Mexico: Landscapes and Popular Sketches*), 154
Scaliger, Joseph, 73
Scaliger, Julius Caesar, 65, 73
Schott, Caspar, 49
Schreck, Johannes, 29
science, definition of early modern, 3
Scientific Revolution, historiography on, 4, 23, 45
scribal culture in Spain, 4, 23
Sempere y Guarinos, Juan, 105, 107
Seniergues, Jean, 62
Sessé, Martín de, 56
Shapin, Steven, 45
Sigüenza y Góngora, Carlos, 48–49
Sloane, Hans, 40
Smith, Adam, 98, 105
Smith, Captain John, 10
Solórzano Pereira, Juan, 84
Soto, J., 133
Soto de Barona, Dolores, 136
Southern Cross, 70–71; and court rituals, 180n35
Spanish Academy of Language, 103
Spanish America: and astrology, 67, 70–83; and Bourbon Reforms, 56–59, 121–22; and corruption of food, 179n23; and Creole degeneration, 74, 75, 78; and Creole patriotism, 50, 62–63, 78, 121–28; institutions of learning, 47–49, 57; and meteorology, 68–69, 117; and paradise, 64, 65, 83, 117–21; and purity of blood, 94; and race, 47, 64–65, 85–95, 140(fig), 163–64; and wars of independence, 113
Spanish Empire: and astronomy, 38; and Black Legend, 23–24, 45, 95; causes of decline of, 24, 97–110, 131; as composite monarchy, 3, 9, 12, 47; and millenarianism, 22, 73; and Siglo de Oro, 32, 103, 110; uses of science in, 7

Index